P9-EMN-972

# PANDORA'S LAB

# PANDORA'S LAB

## Seven Stories of Science Gone Wrong

PAUL A. OFFIT, M.D.

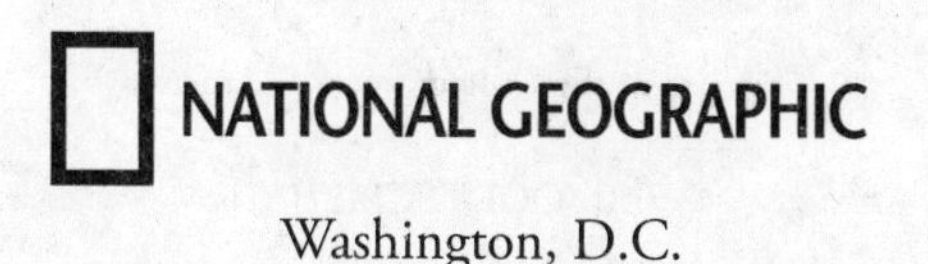

Washington, D.C.

Published by National Geographic Partners, LLC
1145 17th Street NW, Washington, DC 20036



Library of Congress Cataloging-in-Publication Data
Names: Offit, Paul A.
Title: Pandora's lab : seven stories of science gone wrong / Paul A. Offit, M.D.
Description: Washington, D.C. : National Geographic, [2016] | Includes bibliographical references.
Identifiers: LCCN 2016030737 | ISBN 9781426217982 (hardcover : alk. paper)
Subjects: LCSH: Errors, Scientific.
Classification: LCC Q172.5.E77 O34 2016 | DDC 001.9/6--dc23
LC record available at https://lccn.loc.gov/2016030737

Since 1888, the National Geographic Society has funded more than 12,000 research, exploration, and preservation projects around the world. National Geographic Partners distributes a portion of the funds it receives from your purchase to National Geographic Society to support programs including the conservation of animals and their habitats.

National Geographic Partners
1145 17th Street NW
Washington, DC 20036-4688 USA



*Interior design: Katie Olsen*

Printed in the United States of America

16/QGF-LSCML/1

*To my wife, Bonnie, who listened patiently*
*while I talked about rotavirus surface proteins*
*during far too many dinners and vacations;*
*and to our children, Will and Emily,*
*who make everything worthwhile.*

"But [Pandora] took off the great lid of the jar with her hands and scattered all [that] caused sorrow and mischief . . ."

—Hesiod, *Works and Days*

# CONTENTS

# INTRODUCTION

*"Invention does not consist of creating out of void, but out of chaos."*

—Mary Shelley

The Franklin Institute in Philadelphia is home to the Benjamin Franklin National Memorial. Founded in 1824, it's one of the oldest science education centers in the United States. In 2014, the institute featured "101 Inventions That Changed the World." When I visited this exhibition with my son, who is a science writer, we tried to guess which inventions made the list. We got a lot of them right, but some were surprising.

The top three inventions were pasteurization, paper, and controlled fire; rounding out the list were the sail, air-conditioning, and the Global Positioning System (GPS). Among others were the telephone, cloning, the alphabet, penicillin, the spinning wheel, vaccination, transistor radios, email, and aspirin. Two inventions that my son and I would never have predicted were gunpowder (number 20) and the atomic bomb (number 30)—both of which have arguably done far more harm than good. This suggested the possibility of another list: "101 Inventions That Changed the World—For the Worse."

During the past few years, I've asked doctors, scientists, anthropologists, sociologists, psychologists, skeptics, and friends to provide a list of what they think were the world's worst inventions. In the end,

I had about 50 from which to choose. Initially, I thought I'd limit the list to discoveries that had caused the most deaths (like explosives). Then I focused only on those that had harmed the environment (like the refrigerant Freon). In the end, I settled on inventions that were not only the most surprising (at least to me) but also ones whose impact is still felt today.

HERE ARE THE SEVEN FINALISTS:

Six thousand years ago, the Sumerians discovered a plant called *hul gil,* "the plant of joy," which gave birth to a drug that now kills 20,000 Americans every year. More young adults die from this drug than from motor vehicle accidents.

In 1901, a German scientist performed an experiment that revolutionized the food industry. A hundred years later, an editorial in the prestigious *New England Journal of Medicine* stated, "On a per calorie basis, [this product] appears to increase the risk of heart disease more than any other macronutrient." The Harvard School of Public Health estimated that eliminating it from the American diet would prevent 250,000 heart-related deaths every year.

In 1909, another German scientist invented a chemical reaction that won the Nobel Prize, allowed us to feed more than seven billion people across the globe, and unless we do something about it, will probably end life on this planet.

In 1916, a New York City conservationist wrote a scientific treatise that caused Congress to pass a series of draconian immigration laws, enabled the forced sterilization of tens of thousands of American citizens, and provided a scientific rationale for Adolf Hitler to murder six million Jews. Echoes of this treatise can be heard today when politicians like Donald Trump denounce Mexican immigrants, calling them "rapists" and "murderers."

In 1935, a Portuguese neurologist invented a surgical cure for psychiatric disorders that won a Nobel Prize, took only five minutes to perform, caused President John F. Kennedy's sister to be permanently disabled, and is now a subject of horror films. Remnants of this dangerous quick-fix procedure can be found today in promised cures for one of the most common psychiatric disorders of childhood: autism.

In 1962, a popular naturalist—the mother of the modern environmental movement—wrote a book that led to the ban of one particular pesticide. The prohibition was hailed by environmental activists but feared by public health officials. Their fears were well founded; as a consequence of the ban, tens of millions of children died needlessly.

In 1966, with the power of two Nobel Prizes behind him, an American chemist elevated the word "antioxidant" into the pantheon of can't-miss marketing terms. Unfortunately, those who have followed his advice have only increased their risks of cancer and heart disease. Worse, he gave birth to an industry whose harm can be found today in the sudden need for liver transplants in Hawaii or in the strange onset of masculinizing symptoms in women in the Northeast.

All of these stories are united by a myth that dates back to 700 B.C.—the myth of unintended consequences. Zeus, angry that Prometheus had stolen fire from the gods, was intent on punishing mankind. So he gave a marvelous jeweled box to Pandora—its contents, a secret. When Pandora opened the box, which she had been warned not to do, a stream of ghostly creatures representing disease, poverty, misery, sadness, death, and all manner of evil escaped. Pandora closed the box, but too late. Only hope remained.

Science can be Pandora's beautiful box. And our curiosity about what science can offer has allowed us, in some cases, to unleash evils that have caused much suffering and death. In one case, it probably sowed the seeds of our eventual destruction. Because these stories start at the beginning of recorded history and extend to the present day, the lesson of Pandora's box remains unlearned.

As a scientist who has worked on vaccines for the past 35 years, I have witnessed both the joy of science as a panacea and the sadness of unintended consequences. For example, the oral polio vaccine, which eliminated polio from the Western hemisphere and is still used throughout the world, can itself cause polio. Although this side effect is rare, it's real. A rotavirus vaccine given to infants in the United States for ten months between 1998 and 1999 before it was withdrawn was a rare cause of intestinal blockage: One child died as a consequence. A swine flu vaccine given in Europe and Scandinavian countries in 2009 was found to cause a rare but permanent disorder of wakefulness called narcolepsy. All of these inventions were well intentioned, all protected against potentially fatal infections, and all resulted in some level of tragedy.

FOR EACH OF THE SEVEN INVENTIONS that follow, we'll analyze how their deadly outcomes might have been avoided. Then, in the final chapter, we'll apply what we've learned to modern-day discoveries such as e-cigarettes, chemical resins, autism cures, cancer-screening programs, and genetically modified organisms (GMOs) to see if we can distinguish a scientific advance from a scientific tragedy in the making, to see whether we have learned from our past or have once again opened Pandora's box. The conclusions, no doubt, will surprise you.

# CHAPTER 1

# GOD'S OWN MEDICINE

*"Of pain you could wish only one thing: that it should stop. Nothing in the world was so bad as physical pain. In the face of pain there are no heroes."*

—George Orwell, *1984*

The first civilization produced the first medicine.

About 6,000 years ago, around the time of Abraham, the Sumerians migrated from Persia (now Iran) and settled between the Tigris and Euphrates Rivers. They invented cuneiform writing, producing more than 400,000 clay tablets. They invented farming, growing barley, wheat, dates, apples, plums, and grapes. And they discovered a plant that would eventually cause more pleasure and more suffering than any other plant in history. They called it *hul gil,* "the plant of joy." Carl Linnaeus, an 18th-century Swedish botanist, called it *Papaver somniferum.* Today, we call it the opium poppy.

Opium was so powerful that ancient cultures reasoned it could only have come from the gods. The Sumerians believed it was a gift from Isis, who gave it to the sun god, Ra, to treat his headache. In India, enthusiasts believed it had come from Buddha, who had cut off his eyelids to prevent sleep. When the eyelids touched the ground, they became the beautiful flower that would provide sleep and dreams for all eternity. Thomas Sydenham, a 17th-century English physician said, "Among the remedies which it has pleased the Almighty God to give man to relieve his sufferings, none is so universal and efficacious as opium." The characterization of opium as divine persisted into the 20th century. In the early 1900s, William Osler, arguably the most distinguished physician of his day and a founder of Johns Hopkins Hospital, called opium "God's own medicine."

Throughout history, the opium poppy has been accommodating, growing in many different types of soils and locations. It's also naturally resistant to insects and fungi. For these reasons, even countries with limited resources can grow and harvest the plant. (Today, the opium poppy is the leading cash crop of Afghanistan.) The money is in the seedpod, which contains a milky white liquid that hardens into a dark gum. The gum (opium) contains five biologically active ingredients: morphine, the most powerful pain-relieving (analgesic) medicine known to man; codeine (methylmorphine), a mild analgesic and cough suppressant; alpha-narcotine and papaverine, which are muscle relaxants; and thebaine, which, starting in the late 1990s, formed the basis of a drug that is now killing about 20,000 Americans a year.

Since the days of ancient Greece, doctors have used opium to treat pain as well as a vast array of other illnesses.

Hippocrates, the father of modern medicine, used it to treat insomnia. Galen, the last of the great Greek physicians, used it to treat headaches, vertigo, deafness, epilepsy, apoplexy (stroke), poor sight, bronchitis, asthma, coughs, hemoptysis (coughing up blood), colic, jaundice, hardness of the spleen, kidney stones, urinary complaints, fever, dropsy (swelling of the limbs, caused by heart failure), leprosy, menstrual problems, and melancholy. Neither Hippocrates nor Galen was aware of opium's snare. Rather, it was a relatively unknown physician named Diagoras of Melos who was the first to realize that many of his fellow Greeks had become hopelessly addicted to the drug. As a consequence, he became the first person in history to argue against its use, declaring that it was better to suffer pain than to become addicted to opium. His warnings have been ignored for the last 2,500 years.

The Romans were also smitten with the opium poppy, which was emblazoned on their coins and honored by Somnos, their god of sleep. But the Romans also understood that opium could be a powerful poison. In 183 B.C., the Carthaginian general Hannibal used it to kill himself. And the emperor Claudius's wife, Agrippina, used it to poison her 14-year-old stepson, Britannicus, so that her son, Nero, could become emperor.

A reference to opium can even be found in the New Testament. As described in Matthew 27:34, when Jesus was hanging on the cross, his followers offered him something to dull the pain: "They gave him vinegar to drink mingled with gall; and when he had tasted thereof, he would not drink." Because it was bitter, opium was often mixed with wine or beer to make it more palatable. Biblical scholars have theorized that *gall,* which means "something bitter," was probably opium.

Neither the Greeks nor the Romans traded in opium. Rather, opium commerce was the province of Arab merchants, who brought the drug to China—where it enslaved a nation.

Opium first made its way to China in the seventh century A.D., where it was used primarily for medicinal purposes, although sometimes it was added to sweets and cakes. At first, opium was a pleasant distraction. But when the Portuguese brought the smoking pipe to

China, everything changed. Chinese citizens couldn't get enough of the drug.

In 1660, British-owned companies shipped 1,350 pounds of opium from India to China; by 1720, 33,000 pounds; and by 1773, 165,000 pounds. About three million Chinese citizens were addicted. In response, the Chinese government banned opium smoking. It didn't work. By 1839, the British were exporting a startling 5,600,000 pounds. At least 25 percent of the Chinese population was addicted to opium; in some regions, the addiction rate was as high as 90 percent. Chinese society was on the verge of collapse. In response, the Chinese government pleaded with British officials to stop exporting opium from India. When they refused, Chinese officials, desperate to end the massive epidemic of addiction and crime that had overtaken their country, took the next step.

In 1839, Commissioner Lin Tse-hsu seized and destroyed 2,600,000 pounds of British opium. War ensued. Between 1839 and 1860, China and Britain fought two Opium Wars. China lost both times. As a consequence, China had to open more ports for opium importation, pay Britain $21 million in reparations, and cede Hong Kong to British rule (which, by treaty, wasn't returned to China until 1997). Eventually, China legalized the drug. By 1900, China was importing 8,600,000 pounds of opium and had more than 13 million opium addicts.

While the Chinese were smoking opium, Americans—thanks to a European inventor—were drinking it.

IN THE EARLY 16TH CENTURY a Swiss alchemist, physician, astrologer, and philosopher named Paracelsus mixed opium with brandy, calling his concoction laudanum, from the Latin verb *laudare,* meaning, "to be praised." "I possess a secret remedy which I call laudanum and which is superior to all other heroic remedies," he said. Liquid opium swept through Europe. Victorian women, who found it unacceptable to frequent bars and saloons, turned to laudanum; they also gave it to their babies to help them sleep. British physicians used laudanum to treat coughs, diarrhea, dysentery, and gout.

Americans also embraced liquid opium. Louisa May Alcott and George Washington used laudanum; Mary Todd Lincoln was addicted to it. By the late 1800s, about 200,000 opium addicts lived in the United States; three-quarters were women. Unlike opium smokers in China, women in Europe and the United States who drank laudanum were considered to have a gentle, harmless addiction. In Harper Lee's *To Kill a Mockingbird,* set in a small Alabama town, Mrs. Henry Lafayette Dubose is addicted to laudanum, the picture of decay. But Atticus Finch—the lawyer who takes the town to task for its racist beliefs—praises Dubose for her courageous attempt to fight her addiction and die with dignity. Finch sees Dubose as a sympathetic, not pathetic, character.

Opium was also a staple of the patent medicine craze. Medicines like Stott's Unique Fruit Cordial, which contained 3 percent opium, and Chlorodyne, which contained opium, cannabis, and chloroform, were easily purchased over the counter. And Mrs. Winslow's Soothing Syrup, Mother Bailey's Quieting Syrup, and Hooper's Anodyne were all spoon-fed to "quiet the cranky child." The American Medical Association later called opium-containing preparations "baby killers."

The opium poppy also made a cameo appearance in L. Frank Baum's best-selling book, *The Wizard of Oz.*

> [Dorothy's] eyes closed in spite of herself and she forgot where she was and fell among the poppies, fast asleep.
>
> "What shall we do?" asked the Tin Woodsman.

> "If we leave her here she will die," said the Lion. "The smell of the flowers is killing us all. I myself can scarcely keep my eyes open and the dog is asleep already."

UNLIKE THE EUROPEANS, Americans eventually banned opium use. They did it because of a series of events triggered by the California gold rush.

Between 1850 and 1870, about 70,000 Chinese citizens entered the United States to mine gold and work on the railroads, bringing their opium pipes with them. They came through the port of San Francisco. Initially, opium smoking was limited to Chinese immigrants. But, starting in the 1870s, opium dens became a popular haunt for actors, gamblers, prostitutes, and criminals, spreading to almost every major American city, including Los Angeles, New York, Chicago, and Miami. Opium addiction became so widespread and so perverse that in 1875 San Francisco city officials passed the Opium Den Ordinance, prohibiting public smoking of opium. Other cities followed. Then, the United States government stepped in. In 1909, Congress passed the Opium Exclusion Act, banning importation. But it was too late. Many Americans were already addicted to the drug. And, as reflected by a new American lexicon, people addicted to opium were no longer sympathetic figures. They were called *junkies,* because they often sifted through junkyards to find salable items. Or *hop heads,* from the Cantonese phrase *ha peen,* meaning "bird or cow manure."

In 1914, the United States Congress passed the Harrison Act, forcing doctors to register and maintain records of all narcotic prescriptions. (In addition to relieving pain, opium is a narcotic, a word derived from the Greek *narkoun,* meaning "to make numb." All narcotics, by definition, suppress the central nervous system, causing drowsiness, stupor, and occasionally coma.) In 1919, the U.S. Supreme Court extended the act, making it clear that doctors were prohibited from prescribing narcotics to maintain an addiction. Almost a hundred years would pass before doctors were held accountable for violating this law.

Opium was now restricted by state legislatures and reviled by the American public. But the enslavement of Americans by opium and its derivatives was just getting started.

Although opium was clearly addictive to individuals and destructive to society, its pain-relieving properties were undeniable. No other drug could match it. Scientists were desperate to find a way to retain opium's analgesic properties while jettisoning its addictive properties. In the early 1800s, a young German chemist became the first to try.

In 1803, Friedrich Sertürner, a 20-year-old chemist's apprentice, isolated opium's most abundant and most active ingredient. He named it *morphium* after the Greek god of dreams: Morpheus. Later, the name was changed to morphine. Sertürner never trained at a university, never earned a degree, had no professional standing, made his own laboratory equipment, and tested the effects of his newfound product on the only person he could find to do it: himself. A shy and solitary man, this remarkably young chemist's apprentice would soon change the face of medicine.

Sertürner found that morphine was about six times more powerful than opium, causing almost immediate euphoria followed by depression and dependence. When he finished his studies, he was addicted to the drug. Worried that he had created a monster, Sertürner warned, "I consider it my duty to attract attention to the terrible effects of this new substance I called *morphium* in order that calamity may be averted." Sertürner's warning went unheard. By 1827, the German pharmaceutical company Merck began mass-producing the drug. European doctors soon prescribed morphine for a variety of illnesses, including alcoholism, inadvertently shifting the addiction from alcohol to morphine.

Then a medical invention changed the face of narcotic addiction.

In 1853, a doctor in Edinburgh, Scotland, named Alexander Wood attached a syringe to a needle, allowing morphine to be injected directly into the bloodstream. (Morphine was the first intravenous drug.) Wood reasoned that if morphine were injected instead of ingested, people wouldn't develop an "appetite" for the drug. He believed that he had found a way to separate morphine's analgesic properties from its addictive properties. By 1880, almost every physician in the United States owned a hypodermic needle and began instructing patients on how to inject morphine on their own. Wood's wife would later die from a morphine overdose—the first recorded patient to die from an injectable drug.

With the invention of the hypodermic syringe, morphine became the addict's drug of choice. By 1900, more than 300,000 people in the United States were addicted to morphine. When laws were passed to prohibit its sale, the demographics of addiction quickly shifted. No longer were addicts the frail, sympathetic, laudanum-drinking women of *To Kill a Mockingbird;* they were poor urban males like Frankie Machine, the hustling, pool-playing junkie in Nelson Algren's best seller *The Man with the Golden Arm.* (Frank Sinatra played Machine in the 1955 movie.)

It was back to the drawing board. Was it possible to invent a pain-relieving medicine that had the power of opium—and its principal component, morphine—without causing the addiction and dependence that invariably accompanied the drug? At this point, scientists had used only products found in nature. Surely, there must be a way to use modern chemistry to synthesize a nonaddictive painkiller. In the late-1800s, one scientist believed he had discovered it: the holy grail of pain relief.

IN 1874, A PHARMACIST in London named C. R. Alder Wright boiled morphine with the reactive form of acetic acid (vinegar) on a stove for several hours, producing diacetylmorphine. (This process is called acetylation.) Convinced that he had finally created a nonaddictive pain reliever, Wright fed the gray-white powder to his dog, who became frighteningly hyperactive, violently ill, and almost died. After throwing away the powder, he published his findings in the *Journal of the Chemical Society in London*. Although Wright would soon become a Fellow of the prestigious Royal Society, nobody paid attention to what he had written.

Twenty-one years passed.

In the late 1800s, Heinrich Dreser, a young chemistry professor working for a struggling pharmaceutical company in the Rhineland, discovered Alder Wright's article. Like Wright, Dreser wanted to rid morphine of its addictive properties. He was impressed by Wright's paper. Dreser knew that by acetylating morphine, the drug could enter the brain more quickly. Therefore, much less morphine would be required for pain relief. With so little drug required to induce a biological effect, Dreser reasoned that people would be much less likely to become addicted to the drug. At last, a safe, effective pain reliever.

In 1895, Dreser instructed his assistant, Felix Hoffmann, a postdoctoral student, to acetylate morphine. This was nothing new to Hoffmann, who had recently acetylated another chemical, sodium salicylate, which had been used as an anti-inflammatory drug to treat rheumatism. The problem with sodium salicylate was that it damaged the lining of the stomach, causing gastritis, bleeding, and occasionally ulcers. Hoffmann found that by acetylating sodium salicylate, resulting in acetyl salicylic acid, the gastritis problem virtually disappeared. In 1899, Dreser and Hoffmann's company—which was named for its founder, Friedrich Bayer—marketed its new drug, calling it Bayer Aspirin.

Now, Dreser and Hoffmann were ready to see if their success with aspirin would carry over to morphine. So they fed diacetylmorphine to a few rats and rabbits that appeared to love it. Then they fed the gray powder to four workmen in the company who also loved it; indeed, the workers were anxious to repeat the experiment. Then they tried the drug on a few local patients.

In September 1898, Heinrich Dreser presented his findings at the 70th Congress of German Naturalists and Physicians. Dreser claimed that diacetylmorphine could treat colds, sore throats, and headaches, as well as severe respiratory infections like pneumonia and tuberculosis—two leading causes of death. Further, diacetylmorphine was five times more potent than morphine and completely non–habit forming. (At this point, Dreser had tested the drug on only a handful of people for about four weeks.) Dreser believed that he had found the perfect drug to treat morphine addiction. Attendees at the conference gave him a standing ovation.

Dreser didn't have to work hard to convince Bayer executives to launch the new drug. But first they had to come up with a name. Some of the workers wanted to call it *wunderlich,* meaning miracle. But Dreser preferred the name *heroisch,* meaning heroic. In 1898, Bayer launched their new drug, calling it heroin. Aspirin, which physicians worried might cause gastritis, could be obtained by prescription only. Heroin, which was believed to be much safer, could be purchased over the counter.

In 1900, Eli Lilly, working in collaboration with Bayer, began distributing heroin without prescription in the United States, promoting it side by side with aspirin as a treatment for colds and the flu. Lilly claimed that the drug could be given safely not only to children, but also to infants and pregnant women.

Heroin sales took off. First, the military administered the drug intravenously to its soldiers in the field during World War I. Then, citizens bought heroin in the form of cough lozenges or as an elixir mixed in glycerin. Millions of doses were sold in England and the United States. In the early 1900s, the philanthropic Saint James Society launched a campaign to send free heroin to morphine addicts.

Heroin became a standard of care. In 1906, the *Journal of the American Medical Association* stated that heroin was "recommended chiefly for the treatment of diseases of bronchitis, pneumonia, consumption [tuberculosis], asthma, whooping cough, laryngitis, and certain forms of hay fever."

IT DIDN'T TAKE LONG TO REALIZE that heroin wasn't what it was claimed to be.

By 1902, at least a dozen cases of addiction and some infant deaths had been reported. By 1905, the evidence was overwhelming. Because heroin crosses the placenta, infants born to heroin addicts suffered symptoms of severe withdrawal. Traces of heroin could also be found in breast milk. In 1906, the Council on Pharmacy and Chemistry stated, "The habit is readily formed and leads to the most deplorable results." By 1910, doctors were fully aware of the dangers of heroin, and its use declined. Bayer, on the other hand, didn't stop advertising the drug as safe until 1913. By 1918, more than 200,000 people living in New York City alone were addicted to heroin.

In 1924, Congress passed the Heroin Act, making manufacture and sale of the drug illegal. As a consequence, heroin went underground. In the 1920s and early 1930s, heroin's principal distributors were mobsters Meyer Lansky, Dutch Schultz, and Legs Diamond. (Because all three were Jewish, heroin was often called "smack," from the Yiddish word *schmecher,* meaning "addict.") In the mid-1930s the Italian Mafia took over, specifically, Charles "Lucky" Luciano, who established the "French connection." Opium grown in French Indochina or Turkey was shipped to Lebanon where it was converted to morphine and then shipped to the French port city of Marseille where it was processed into high-quality heroin and smuggled into the United States.

Initially, heroin abuse was confined to a poor, urban underclass. By the 1940s, however, heroin addiction had spread to the Harlem jazz scene, and by the 1950s—through the writings of Jack Kerouac and William Burroughs—to the beat generation. By the mid-1960s, more than 500,000 Americans were addicted to heroin. Virtually all major U.S. cities, as well as countries like Britain, France, and Germany, were caught in heroin's snare.

The U.S. government took action, pressuring Turkey to stop producing opium and eliminating importation of heroin from France. (This success was dramatized in the 1971 movie *The French Connection,* starring Gene Hackman and Roy Scheider.)

By the 1970s, opium production had moved to an area in the highlands of Laos, Thailand, and Burma (now Myanmar), known as the Golden Triangle. No group suffered this switch in opium production more than American soldiers in Vietnam, about 15 percent of which became addicted to heroin.

In the summer of 1971, President Nixon declared an "all-out war on drugs." "America has the largest number of heroin addicts of any nation in the world," he said. "If we cannot destroy the drug menace in America, then it will surely in time destroy us." Nixon chose Elvis Presley to be the public face of his war on drugs. Ironic, given that at the time of Presley's death in 1977, Valium, methaqualone, morphine, codeine, and barbiturates were found in his bloodstream. Presley wasn't the only celebrity to die from a drug overdose: Janis Joplin died in 1970, John Belushi in 1982, Chris Farley in 1997, and Philip Seymour Hoffman in 2014, all from heroin overdoses.

By the mid-1990s, heroin—which had become cheaper and purer—could be liquefied on tinfoil and its vapors inhaled (called "chasing the dragon"). Now more women started to use the drug. By 1995, more than 600,000 Americans were addicted to heroin. In addition to the Golden Triangle, the Medellin cartel in Colombia also produced large quantities of the drug. The Drug Enforcement Agency, which now had more than 75 offices in 50 countries, was spending more than $13 billion a year trying to keep heroin out of the country.

By 2003, the number of Americans addicted to heroin had decreased from 600,000 to a little more than 100,000. This decrease wasn't because Americans had lost interest in narcotics. It was because they had again replaced one addiction with another.

SCIENTISTS HAD HOPED that morphine could treat opium addiction. Then they had hoped that heroin could treat morphine addiction. It was time to try something else. Again, they would synthetically modify a drug to separate pain relief from addiction. And again, they would fail. This time, spectacularly.

To find the next wonder drug, scientists turned to another component of opium: thebaine, named for Thebes, a town in ancient Egypt where the opium poppy was grown. The first synthetic version of thebaine was produced in 1916 by two German chemists working at the University of Frankfurt. They called it oxycodone.

In the early 1950s, oxycodone made its American debut. Initial preparations were combined with a variety of other drugs. For example, there was Percodan, a combination of oxycodone and aspirin; Combunox, a combination of oxycodone and ibuprofen, a nonsteroidal anti-inflammatory; and Percocet, a combination of oxycodone and acetaminophen (Tylenol). But the single most powerful, and eventually most addictive and most abused preparation, was OxyContin, pure oxycodone uncut by other drugs. OxyContin's manufacturer, Purdue Pharma, marketed the drug as a first-line agent for arthritis. In OxyContin, Purdue Pharma had struck gold—the drug would eventually account for more than 80 percent of its business.

Later, Purdue combined OxyContin with an acrylic that allowed for a slower, timed release of the drug, eliminating its need to be taken several times a day. Addicts soon found that by chewing the tablet or crushing it, they could bypass the timed-release mechanism and enjoy the immediate rush of as much as 160 milligrams of oxycodone, logarithmically greater than any other product on the market. Now addicts had the capacity to ingest a potentially lethal dose of the drug. (On a weight-by-weight basis oxycodone is actually *more* powerful than morphine.)

When OxyContin first came onto the market in 1996, the label read, "Delayed absorption, as provided by OxyContin tablets, is believed to reduce the abuse liability of the drug." Officials from the Food and Drug Administration (FDA) would soon regret this label. In the end, there was nothing controlled about the controlled-release form of OxyContin.

Physicians were initially wary of oxycodone. They had been burned by morphine in the 1800s, heroin in the 1900s, and opium since the beginning of recorded history. They didn't want to be burned again. So they were slow to prescribe the next opium-derived miracle. By the mid-1980s, however, all of that would change.

On April 20, 1948, Cicely Saunders, a nurse, joined St. Luke's Hospital for the Dying in East London. Saunders believed that patients with terminal illness shouldn't have to spend their last few weeks crying out in pain. Rather, they should die a dignified death—one as pain-free as possible. Saunders reasoned that it was better to prevent pain than to treat it. So, in 1967, she founded the hospice movement, providing dying patients with large quantities of addictive, pain-relieving medicines. Saunders's movement crossed the ocean. In 1984, the United States Congress passed the Compassionate Pain Relief Act, making it legal to treat terminally ill patients with heroin. In 1986, Wisconsin launched the first state-based pain management program for cancer patients. Other states followed.

For many patients suffering from terminal illnesses, vigorous pain management was a godsend. But a door had now been opened for doctors to prescribe long-term, high-dose narcotics. Initially, use was limited to patients with terminal cancer. Then a respected doctor from New York City took the liberal use of narcotics one ill-fated step farther.

In 1986, Russell Portenoy, a 31-year-old New York City pain specialist, published a paper in the journal *Pain*. Portenoy believed it was time for American physicians to get over their fear of painkillers, what he called "opiophobia." Portenoy reported the stories of 38 people who were on high-dose pain medicines (12 were on OxyContin). Only 2 of the 38 had become addicted to their drugs, and both had had a history of addiction. Portenoy argued that his findings weren't unique; three previously published studies had also shown that less than one percent of patients on chronic painkillers had become addicted to them. Portenoy reasoned that, "opioid maintenance therapy can be a safe, salutary and more humane alternative [for] patients with intractable non-malignant pain and no history of drug abuse." Russell Portenoy believed that the compassion Cicely Saunders showed for patients with terminal cancer should be extended to all patients. Pain, argued Portenoy, should now be the fifth vital sign (in addition to temperature, blood pressure, heart rate, and respiratory rate). No one should be allowed to suffer. (One note on nomenclature: When Russell Portenoy used the term *opioid,* he was referring to synthetic forms of opium, like oxycodone. Morphine and codeine, which can be purified directly from opium without modification, are called *opiates.*)

Charismatic, bright, and persuasive, Russell Portenoy became the media's "go-to" guy for pain management, frequently appearing in newspapers and popular magazines. His academic success was also meteoric; Portenoy wrote or co-wrote more than 140 papers in medical and scientific journals and 15 book chapters. When Russell Portenoy talked, doctors listened. Portenoy had now given doctors permission to come back to opium derivatives. This time, he assured them, there would be little addiction and death. The days of opium, morphine, and heroin were behind them. Drugs like oxycodone had finally solved the problem of pain relief without addiction. Richard Nixon's war on drugs had become Russell Portenoy's war on the war on drugs.

In late 1995, at the same time that Russell Portenoy was urging American physicians to get over their fear of painkillers, the FDA approved Purdue Pharma's timed-released version of OxyContin. Purdue's sales force promoted the drug for the treatment of lower back pain, arthritis, trauma, fibromyalgia, dental procedures, broken bones, sports injuries, and pain resulting from surgery. In other words: everything. And they constantly repeated Portenoy's mantra that less than one percent of patients would become addicted to the drug.

In 1996, more than 300,000 prescriptions for OxyContin were written for a net profit to Purdue Pharma of $44 million. Realizing it had created the right drug at the right time, Purdue doubled its sales force, offered coupons for a free 7- to 30-day supply of the drug (34,000 coupons were redeemed), increased its yearly advertising budget to $200 million, and paid $40 million in incentive bonuses. In 2001, Purdue reaped $1.45 billion from the sale of OxyContin, the highest retail sale of any named pharmaceutical product, including Viagra.

SALES OF OXYCONTIN also benefited from a vigorous black market.

More than 70 percent of recreational OxyContin users had procured the drug from friends or relatives; 5 percent from drug dealers on the Internet. Sometimes users stole the drug from pharmacies; 90 percent of robberies in Pulaski, Virginia, were due to OxyContin abuse, and half of the inmates in Hazard, Kentucky, were incarcerated for OxyContin-related crimes. Sometimes, to supplement their meager Social Security checks, the elderly poor used Medicare or Medicaid to acquire a bottle of a hundred 80-milligram tablets of OxyContin and sold them on the street for $1.00 per milligram, netting an $8,000 profit for the seller. Teenagers stole OxyContin from their parents. (One street name for OxyContin was "kiddie dope.") Prescriptions were forged. Women turned to prostitution to feed their habit. Pharmacists diverted the drug and sold it on the side. Before he was arrested, one Pennsylvania pharmacist had illegally sold hundreds of thousands of dollars worth of prescription painkillers, mainly OxyContin, over a three-year period, netting him $900,000 (which he later lost in the stock market).

Doctors took advantage of the OxyContin gold rush, selling prescriptions for money or sex. Dr. Randolph W. Lievertz of Indianapolis wrote more than $1 million in prescriptions paid for by the state's Medicaid program; $130,000 of that total was written for one female patient who was part of a drug ring that sold OxyContin on the street. To honor the prescription, the woman would have had to ingest 31 tablets every 12 hours instead of the one tablet recommended by the manufacturer. Lievertz wasn't alone. Pill mills sprang up across the country. One eastern Kentucky doctor saw 150 patients a day, writing prescriptions for painkillers after spending less than three minutes with each patient. Florida alone was home to hundreds of these facilities.

Doctors were arrested and charged with manslaughter and murder. Some were jailed. No case, however, drew more national media attention than that of Dr. James Graves, a 55-year-old Florida physician who was charged with manslaughter in the overdose deaths of four of his patients. Graves's prescription mill was widely known among addicts. "The word spread that he was the go-to doctor to get pills," said Russ Edgar, the assistant state attorney. Edgar argued in court that Graves had bragged that writing prescriptions for painkillers was a "gold mine" because he had rarely examined patients and didn't have to fill out medical records. Edgar also claimed that Graves had ignored the pleas of pharmacists and parents to change his prescribing habits. Indeed, Graves's parking lot often resembled scenes more commonly found at sporting events. Patients would eat, work on their cars, and give each other high-fives when they emerged with their OxyContin prescriptions. "You've got to realize something's wrong when outside your office people are having tailgate parties," said Edgar.

During the trial, Edgar argued that, "Mother after mother after mother called the defendant's office and asked him to quit giving their children drugs or they would die. The defendant did not quit and they continued to overdose." Graves countered that no one would have died if patients had simply used the drugs as prescribed. And he railed against the prosecutor, whom he considered ungodly, telling the judge, "I pray to God something will change and somehow he will come to know Christ." James Graves was sentenced to 63 years in prison for manslaughter. He was the first physician to be found guilty of manslaughter or murder in connection with the irresponsible prescribing of a painkiller.

NO AREAS SUFFERED THE NIGHTMARE of OxyContin pill mills more than rural Appalachia and the Ohio Valley.

OxyContin abuse first surfaced in rural Maine in the late 1990s, then spread down the East Coast to include West Virginia, Kentucky, and southern Ohio. (Another name for OxyContin was "hillbilly heroin.") From 1995 to 2001, the number of patients treated for oxycodone abuse in Maine increased 460 percent, and in eastern Kentucky, 500 percent. In West Virginia, six new drug treatment clinics treated more than 3,000 addicts. Southwestern Virginia opened its first drug treatment center in 2000; within three years more than 1,400 patients were being treated; by 2003, the region had experienced an 830 percent increase in deaths from oxycodone abuse. By 1999, deaths from oxycodone use in Allegheny County in western Pennsylvania outnumbered car fatalities.

Appalachian emergency department physicians became experts in recognizing the symptoms of drug withdrawal, which included anxiety, runny nose, sweating, yawning, insomnia, loss of appetite, gooseflesh (the source of the phrase "going cold turkey"), back pain, abdominal pain, tremors, and occasional involuntary thrusting of the legs (the source of the phrase "kicking the habit.")

IN 2003, 17 YEARS AFTER Russell Portenoy published his paper claiming that long-term use of oxycodone was relatively harmless and nonaddictive, Jane Ballantyne published a paper in the *New England Journal of Medicine* claiming exactly the opposite. Ballantyne showed that long-term use of drugs like OxyContin induced tolerance (larger and larger doses of the drug were required to induce the same effect), hyperesthesia (pain experienced while using painkillers was actually *worse* than the original pain), hormonal changes (specifically, a decline in production of cortisol, an important regulatory hormone), changes in the immune system, and a reduction in fertility, libido, and sex drive. Ballantyne concluded, "Whereas it was previously thought that unlimited dose escalation was at least safe, evidence now suggests that prolonged, high-dose opioid therapy may be neither safe nor effective."

Ballantyne's paper wasn't news to the FDA, which had already changed the label on OxyContin. No longer did the label state that the timed-release formulation made it *less* likely for abuse; now it stated that the formulation made it *more* likely for abuse, addiction, overdose, and death. The warning wasn't written in fine print; rather, the FDA issued its most strident alert—the so-called "black box" warning.

But it was too little, too late.

In 2002, a survey at a rural Michigan high school showed that 98 percent of students had heard of OxyContin and 9.5 percent had tried it; of those who had tried it, 50 percent had taken it more than 20 times. By April, 1,300 deaths caused by OxyContin had been reported to the FDA; in most cases, doctors had prescribed the drug. By the end of 2002, Purdue Pharma was selling more than $30 million worth of OxyContin every week, and sales had exceeded more than $2 billion a year.

In 2003, Rush Limbaugh, a conservative radio commentator who often railed against drug abusers for being morally bankrupt, admitted that he was addicted to OxyContin.

In 2004, three million people were using OxyContin—now the most prevalent prescription painkiller in the United States.

In 2007, 14,000 people died from overdoses of prescription painkillers, and health care and criminal justice system costs exceeded $55 billion.

In 2008, 15,000 people died from prescription painkillers—the leading cause of accidental death in 30 states.

In 2009, health insurers spent $72 billion in direct health care costs related to prescription painkillers.

By 2010, 22 million people had misused prescription painkillers, and more people had died from these drugs than from heroin and cocaine combined. Enough painkillers were now being prescribed to medicate every adult living in the United States around the clock for a month.

In 2012, 12 million Americans aged 12 and older reported the recreational use of prescription painkillers; 16,000 of those users died from overdoses. Painkillers were now the most widely prescribed class of drugs in the United States; every 19 minutes someone died from an overdose. (OxyContin overdoses were indistinguishable from overdoses of opium, morphine, or heroin, all of which suppress the rate and depth of respiration. Patients breathe as few as four times a minute; blood pressure begins to drop, body temperature falls, and the skin becomes cold and clammy. Because the brain isn't getting enough oxygen, the patient seizes and eventually dies from respiratory failure.)

In 2014, U.S. retail pharmacies dispensed 245 million prescriptions for opioid pain relievers. About 2.5 million adults were addicted to the drug.

Few insurance companies did much to discourage abuse. Before OxyContin burst onto the scene, chronic pain was treated with a combination of psychotherapy, biofeedback, exercise, and physical therapy. The goal was to leave the painkillers at the door. Although several studies had showed that this multidisciplinary approach to relieving pain worked just as well if not better than chronic drug use, the fact remained that the drugs were less expensive than the therapies. Unfortunately, many insurance companies encouraged the drugs. At best, this was shortsighted. Workers who took high doses of painkillers were out of work three times longer than those with similar injuries who took lower doses of the drugs.

On May 10, 2007, Purdue Pharma, along with three company executives, pleaded guilty to one count of "misbranding" OxyContin. The court ruled that Purdue had minimized risks, made unsubstantiated claims, and failed to include clear warnings about how, under certain conditions, the drug could be fatal. Just as Bayer had continued to market heroin when it was evident that the drug was causing harm, Purdue had been slow to inform the public about the potential dangers of OxyContin. The three executives were fined $34.5 million (which Purdue paid), barred from working for any company that sold medical products for the next 12 years, and required to perform 400 hours of community service in drug-treatment centers. Purdue also paid an additional $634 million penalty. Many of the parents whose children had died from OxyContin were present during the sentencing. "Our children were not drug addicts; they were typical teenagers," said one woman, whose 19-year-old son had died. "We've been given a life sentence." Judge James P. Jones, who presided over the trial, said he would have liked to have sent company executives to prison, but was bound by the plea bargain.

In August 2010, Purdue replaced their timed-release form of OxyContin with a "tamper-resistant" product. The new version formed a thick, sticky gel that made it more difficult to crush. Two years later, a study in the *New England Journal of Medicine* examined the impact of this new formulation. Researchers found that although OxyContin abuse had decreased, 24 percent of users had found a way to defeat the tamper-resistant properties and 66 percent had just switched to another drug, mainly heroin. Despite the reformulation, yearly sales of OxyContin still topped $2 billion.

In 2012, Russell Portenoy, who had become chairman of pain medicine and palliative care at Beth Israel Medical Center in New York City, recanted: "I gave innumerable lectures in the late 1980s and '90s about addiction that weren't true," he said. "We didn't know then what we know now." During the previous decade, more than 100,000 people had died from overdoses of painkillers. Steven Passik, a psychiatrist who had worked closely with Portenoy, recalled the war on pain: "It had all the makings of a religious movement," he said. "It had a kind of spirit to it."

In the end, OxyContin was one of the most addictive narcotics ever sold. And Russell Portenoy's war on pain was one of modern medicine's biggest mistakes.

On January 16, 2016, an article written by Gina Kolata and Sarah Cohen in the *New York Times* stated: "The rising death rates for young white adults make them the first generation since the Vietnam War years of the mid-1960s to experience higher death rates in early adulthood than the generation that preceded it." Opioid overdoses were now the leading cause of accidental deaths in the United States.

On March 15, 2016, the Centers for Disease Control and Prevention (CDC) finally offered guidelines for the sensible use of prescription painkillers, stating that doctors should prescribe them: (1) only after nonprescription painkillers like ibuprofen and physical therapies had failed; (2) in quantities not to exceed a three-day supply for short-term pain and rarely longer than seven days (typically, patients are given two weeks or a month worth of pills); and (3) only when improvement was significant. The guidelines did not apply to patients who were receiving painkillers for cancer or end-of-life treatment. The American Academy of Pain Management—a group that receives funds from Purdue and Teva Pharmaceuticals—and the Washington Legal Foundation, which frequently represents pharmaceutical company interests in court, opposed the new guidelines. After all, painkillers were now a $9 billion a year industry. Robert Twillman, executive director of the academy, didn't like the three- to seven-day dosing recommendation. "The numbers are still arbitrary," he said. But Tom Frieden, director of the CDC, had had enough. "For the vast majority of patients with chronic pain," he countered, "the known, serious, and far too often fatal risks far outweigh the transient benefits. We lose sight of the fact that prescription opioids are just as addictive as heroin." Today, 80 percent of the world's opioid prescriptions are written in the United States, even though only 5 percent of the world's population lives there.

The lesson from the ill-fated war on pain is a simple one: **It's all about the data**. When Friedrich Sertürner feared that in morphine he had opened up a Pandora's box and let loose a monster, his warnings were ignored. When Heinrich Dreser claimed that heroin was safe, he had only tested it on a handful of people for a few weeks. And when Russell Portenoy launched a national campaign promoting opioid use, he based his claims on 38 patients, 12 of whom had taken OxyContin. As Clara Peller said in her now iconic television commercial for Wendy's hamburgers, "Where's the beef?" If you're going to medicate a nation, at the very least you should base your recommendations on a mountain of evidence, not a molehill.

## CHAPTER 2

# THE GREAT MARGARINE MISTAKE

*"Avoid fried foods, which angry up the blood."*

—Satchel Paige

I grew up in Baltimore, Maryland. And although I know that you cannot go home again—that you can never recapture your youth—still, there were certain things that I had assumed would always be there.

For example, from the ages of 13 to 18, I was a Baltimore Colts season ticket holder. Every year, three friends (Jimmy, Jack, and Robert) and I would scrape up the $35 it took to buy a season ticket for the seven home games. On Sundays, we would take the bus down to Memorial Stadium: "the world's largest outdoor insane asylum." Baltimore loved their Colts. We couldn't have supported them more. Then, with little warning, they were gone—off to Indianapolis in the middle of the night with their memorabilia crammed into the back of a Mayflower truck. Loyalty, apparently, could be a one-way street.

Another Baltimore staple were crabs from the Chesapeake Bay. During the summer we would go to a local crab house and crack crabs flavored with Old Bay Seasoning made by Baltimore's own McCormick & Company. Then, due to overfishing, the crabs were gone. Now Baltimore's crabs are flown up from Texas.

There was another tradition that I could always count on—something that won the "Best of Baltimore Award" in 2011 and has been featured on *The View* as well as the Food Network's *Rachael Ray* and *The Best Thing I Ever Ate*. Something that every Baltimorean adores: Berger cookies.

Slathered with fudge over a modicum of shortbread, Berger cookies have been a Baltimore tradition since George and Henry Berger brought them over from Germany in 1835. Today, Berger cookies are made by a little bakery in the Cherry Hill section of Baltimore that, in 2012, had annual sales of $2.5 million, 98 percent of which was from Berger cookies. Amazing, when you consider that most of these cookies are distributed locally.

Unfortunately, like the Baltimore Colts and Chesapeake Bay crabs, Baltimore's Berger cookies might also soon be a thing of the past. Unless the owner of the bakery, Charles DeBaufre, Jr., changes his recipe, the FDA will ban them. The reason: Berger cookies are made with partially hydrogenated vegetable oils, so they're loaded with trans fats. DeBaufre has tried baking them with cooking oils and shortenings free of trans fats, without success. "We've tried it and trust me, it's nasty," he says. "The texture's just not there. It's an entirely different product." If DeBaufre cannot come up with an alternative recipe soon, his cookie and his business will be gone.

THE THREAT TO BALTIMORE'S Berger cookies is explained by the most common reason that Americans die: heart disease, a phenomenon of the modern era. In the early 1900s, most people died from bacterial and viral infections. But, during the 20th century, advances like antibiotics, vaccines, safer drinking water, and purer foods have allowed us to live about 30 years longer—long enough to die from heart disease. To understand why, we first need to understand what makes the heart so vulnerable.

The heart is a muscle that, like any other muscle, needs the constant flow of blood, which supplies oxygen. Two major arteries, called coronary arteries, do this. If either of these arteries is blocked, then blood flow is disrupted, causing damage to the heart muscle and occasionally sudden death (that is, a heart attack). When researchers studied these blockages, they found cholesterol, a substance made by the body that is an essential component of cell membranes. They also found triglycerides, the main constituent of body fat. Doctors would eventually call the disease atherosclerosis, literally meaning "hardening of the arteries."

The next question was what, if anything, could be done about it. In 1913, Nikolay Anichkov offered the first ray of hope. Working in the Czar's Military Medicine Institute in St. Petersburg, Russia, Anichkov found that rabbits fed large quantities of milk and egg yolks—foods rich in cholesterol—developed atherosclerosis. He reasoned that heart disease could be controlled by diet. Eat less cholesterol, he said, and you'll live longer.

By the mid-1950s, Ancel Keys argued that cholesterol wasn't the only problem. Keys studied people's diets in seven different countries. He found that residents of Japan and Crete had very little heart disease, while those living in Finland—where the amount of fat in the diet was greater—suffered a higher incidence. He urged Americans to restrict their fat intake, becoming the first person to use the term "heart-healthy diet." Despite the clarity of his recommendation, Keys admitted that "direct evidence on the effect of diet on human atherosclerosis is very little and likely to remain so for some time."

Unlike Anichkov, whose work had little influence, Keys had clout. He chaired the International Society of Cardiology for the World Health Organization, was consultant to the United Nations' Food and Agriculture Organization, and, along with his wife, wrote several best-selling books on diet and disease. In 1961, Ancel Keys appeared on the cover of *Time* magazine urging Americans to eat less fat and less cholesterol. That same year, the American Heart Association set a recommended limit of 300 milligrams of dietary cholesterol a day. Because a single egg contains about 200 milligrams of cholesterol, egg consumption dropped 30 percent. "In America, we no longer fear God," said David Kritchevsky, a scientist at the Wistar Institute in Philadelphia. "We fear fat."

Although scientific data on the relationship between fat consumption and human health remained, at best, ambiguous, the United States federal government was determined to impose clarity. In 1968, Senator George McGovern launched the Senate Select Committee on Nutrition and Human Needs. McGovern and his wife had recently tried diet guru Nathan Pritikin's low-fat diet and exercise program. Although McGovern had bailed out on the diet quickly, he remained committed to its mantra.

In 1977, McGovern's committee published its unprecedented and, according to one historian, "revolutionary" report. What made it revolutionary was that it was written by a group of political activists with no specific training or expertise in the field of nutrition. The author of the report was Nick Mottern, a labor reporter for the *Providence Journal*. Mottern had no background in science, nutrition, or human health. So, he turned to the one man he believed could help him decide what diet was right for the American public: Mark Hegsted, a Harvard School of Public Health nutritionist who unconditionally embraced the benefits of restricting dietary fats, even though he admitted that it was an extreme position. Mottern's report, titled "Dietary Goals for the United States," stated that Americans should cut their total fat intake to less than 30 percent of total calories.

The McGovern committee guidelines would have quietly died the death they deserved had it not been for Carol Tucker Foreman, a consumer activist who had recently been appointed U.S. Department of Agriculture assistant secretary of Food and Consumer Services. Foreman decided to elevate the committee's recommendations to official government policy. Undeterred by the lack of clarity from scientific studies, Foreman marshaled forth. "I have to eat and feed my children three times a day," she told a group of scientists, "and I want you to tell me what your best sense of the data is right now." Unfortunately, the "best sense of the data" depended on whom you asked. Scientists just didn't know enough to make a clear recommendation. But the USDA recommendations were clear, even if the data weren't. Restriction of dietary fat became official government policy.

After Mottern's report was made public, McGovern's staffers decided that it might be a good idea to get input from more than one scientist. So, they opened up their committee hearings to others. One of the first to appear was Robert Levy, a senior scientist from the National Heart, Lung, and Blood Institute. Levy testified that no one really knew whether lowering cholesterol or fat intake would do anything to prevent heart disease, and that his institute was in the midst of a $300 million study to find out. But Levy also knew that the horse was out of the barn. "The good senators came out with the guidelines and then called on us to get advice," he lamented.

Next to argue against the committee's report was Pete Ahrens, a metabolism researcher at New York's famous Rockefeller Institute, who, in 1969, had headed a committee that came to the same conclusions as Robert Levy. Even the American Medical Association weighed in, protesting that the diet proposed by McGovern's committee had the "potential for harmful effects." But it was too late. According to Gary Taubes, in a *Science* magazine article titled, "The Soft Science of Dietary Fat," "It was George McGovern's [Committee]—and, to be precise, a handful of McGovern's staff members—that almost single-handedly changed nutritional policy in this country and initiated the process of turning the dietary fat hypothesis into dogma." Although they didn't know it at the time, Americans were now unwitting test subjects in a national experiment to see if restricting dietary fat reduced the incidence of heart disease.

PERHAPS NO PRODUCT SUFFERED government restrictions more than butter, whose origins date back to the time that humans domesticated animals, about 10,000 years ago. Butter is made by separating cream from milk and churning it into a solid, which is naturally a light yellow color. When Keys and McGovern made their definitive, if ill-founded, recommendations, they caused Americans to prefer a product that was first commissioned in 1869 by Napoleon III of France. Napoleon needed something cheaper than butter to feed his army. The first to step forward was a French chemist named Hippolyte Mège-Mouriès, who invented something he called oleomargarine. Unlike butter, which was made from animal fat, margarine was made from vegetable oils. Also unlike butter, margarine was stark white, not light yellow. Cheaper, but similar in taste and texture to butter, margarine soon became one of the most popular food products in the world.

In 1886, the National Dairy Association in the United States fought back, influencing the federal government to pass the Oleomargarine Act, which imposed a tax on anyone selling margarine. To avoid the tax, some margarine makers dyed their product yellow and sold it as butter. Incensed, the dairy industry used its influence to prohibit margarine makers from dying their product. Manufacturers responded by selling margarine with a yellow dye on the side. If consumers wanted their margarine to be yellow, all they had to do was put it in a bowl and add the dye themselves. Three states—Vermont, New Hampshire, and West Virginia—went one step further, passing laws that margarine had to be dyed pink. The margarine tax laws were repealed in 1950, the dye laws in 1955. (Major dairy states like Minnesota and Wisconsin didn't repeal their dye laws until 1967.) Now, margarine could be sold as a yellow, spreadable product free of a federal tax. Advertisers were quick to promote its benefits over butter.

In 1911, the average American ate about 19 pounds of butter a year compared with only 1 pound of margarine. By 1957, with margarine now being offered as the "heart-healthy" alternative, Americans were eating 8.5 pounds of margarine a year, about the same as butter. "The massive advertising of health claims for margarine transformed a generally disreputable product of inferior quality and flavor into a great commercial success," wrote William Rothstein, in his book *Public Health and the Risk Factor*. Even Eleanor Roosevelt jumped in. "That's what I spread on my toast," she said in a 1959 television commercial for Good Luck margarine. By 1976, margarine consumption had increased to 12 pounds a year, three times that of butter. But despite the switch from butter to the supposedly "heart-healthy" margarine, the incidence of heart disease in the United States continued to rise. It took decades for policy makers to understand why margarine was actually the "heart-unhealthy" alternative.

During the next 20 years, three major studies involving 300,000 people and costing about $100 million determined the relationship between dietary fat and heart disease. The answer: There wasn't any. Nonetheless, despite the clarity of these studies, official government policy remained unchanged. Walter Willett, a Harvard epidemiologist who had headed one of the studies, was incensed. "Scandalous," he remarked. "They say, 'You really need a high level of proof to change the recommendations,' which is ironic because they never had a high level of proof to set them."

Ancel Keys and the McGovern committee had been wrong about dietary fats because they had assumed that all fats were the same. They hadn't accounted for the different types of fats, specifically, saturated fats, unsaturated fats, cis fats, and—most important—trans fats. In the years that followed, Americans would pay a high price for their ignorance.

To UNDERSTAND WHERE Keys and McGovern had gone wrong, we're going to need a brief refresher course in high school chemistry for the few people who might have forgotten it. Just kidding. Everyone's forgotten it. You forget it the minute the test is over. But to understand what words like "saturated" and "unsaturated" and "trans fats" mean, we need to understand some of the chemistry behind them. It's really not that hard. So hang in there.

Fats are composed of three different types of atoms: carbon (C), hydrogen (H), and oxygen (O). Carbon atoms, which form the backbone of fats, have four binding sites (areas where one atom attaches to another). If all four sites are bound, then the carbon atom's binding sites are said to be *saturated.* The fat shown below is a saturated fat. Foods rich in saturated fats include butter, lard, coconut oil, palm oil, mayonnaise, and fish oils; dairy products like cream, cheeses, milk, sour cream, and ice cream; and processed meats like bacon, sausage, salami, steak, ham, ground beef, and luncheon meats.

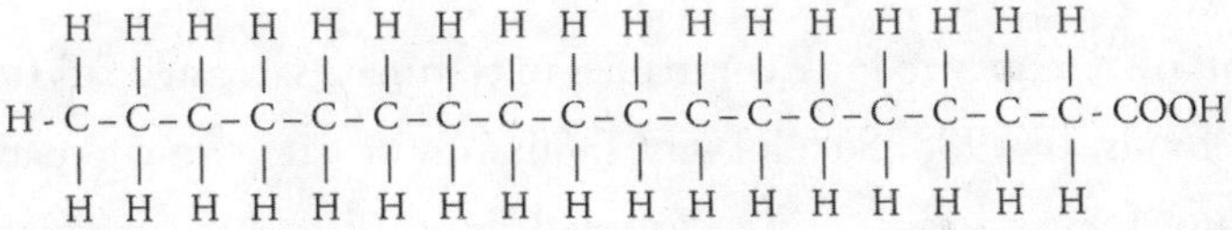

```
    H   H   H   H   H   H   H   H   H   H   H   H   H   H   H   H
    |   |   |   |   |   |   |   |   |   |   |   |   |   |   |   |
H - C - C - C - C - C - C - C - C - C - C - C - C - C - C - C - C - COOH
    |   |   |   |   |   |   |   |   |   |   |   |   |   |   |   |
    H   H   H   H   H   H   H   H   H   H   H   H   H   H   H   H
```

Saturated fat

Sometimes, however, a carbon atom will share *two* binding sites with another carbon atom (such as in the carbon atoms pictured in bold in the example below). Because it's still possible for these carbon atoms to share one of their binding sites with another atom (like a hydrogen atom), the fat is said to be *unsaturated.* The fat shown below is an unsaturated fat. Foods rich in unsaturated fats include olive oil, salmon, almonds, walnuts, pistachios, avocados, olives, fatty fish, margarine, natural peanut butter, and pumpkin, sunflower, flax, and chia seeds.

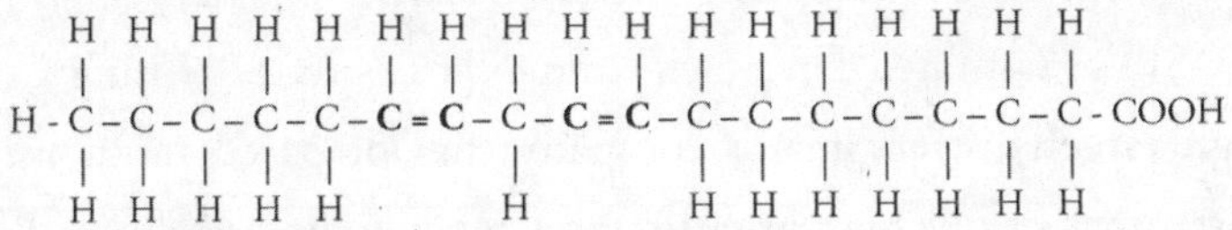

```
    H   H   H   H   H   H   H   H   H   H   H   H   H   H   H   H   H
    |   |   |   |   |   |   |   |   |   |   |   |   |   |   |   |   |
H - C - C - C - C - C - C = C - C - C = C - C - C - C - C - C - C - C - COOH
    |   |   |   |   |           |           |   |   |   |   |   |   |
    H   H   H   H   H           H           H   H   H   H   H   H   H
```

Unsaturated Fat

By the early 1980s, when the relative quantities of these two different types of fats were clear, several studies had shown that saturated fats increased the risk of heart disease. These studies gave birth to the notion that unsaturated fats were good and saturated fats were evil. In response, two groups made it their mission to eliminate saturated fats from the American diet. It wasn't until much later that Americans realized what they'd done wrong.

In 1984, the Center for Science in the Public Interest (CSPI) launched its "saturated fat attack," targeting companies that fried or baked foods using animal fats and tropical oils that were rich in saturated fats (like coconut oil and palm oil). A year later, Phil Sokolof—after suffering a near fatal heart attack—launched the National Heart Savers Association (NHSA), spending $15 million of his own money to force companies to eliminate saturated fats from fast foods. In 1988, Sokolof sent thousands of letters to companies urging them to stop using saturated fats. When his letters were ignored, he took out full-page ads in the *New York Times, Washington Post, New York Post, USA Today, Wall Street Journal,* and other newspapers. "Who is poisoning America?" his advertisements blared, "Food processors are by using saturated fats!" The text that followed was no less subtle. "We have contacted all of the major food processors beseeching them to stop using these potentially dangerous ingredients . . . Our pleas have gone unanswered. Obviously these companies have more pressing priorities than your health. SOMETHING MUST BE DONE . . . We implore you. Do not buy products containing coconut or palm oil. YOUR LIFE MAY BE AT STAKE."

CSPI's "saturated fat attack" and NHSA's letter-writing campaign targeted every major company that prepared foods using shortenings or oils high in saturated fats, including Archway, Borden, Frito-Lay, General Foods, Hardee's, Heinz, Hostess, Keebler, Kellogg's, Kentucky Fried Chicken, Lance, McDonald's, McKee Baking Company, Nabisco, Pepperidge Farm, Pillsbury, Procter & Gamble, Quaker Oats, Ralston Purina, Roman Meal, Roy Rogers, Specialty Bakers, Stouffer's, Sunshine, Taco Bell, and Wendy's. By the late 1980s, virtually every major cookbook and every reputable dietitian promoted diets low in saturated fats, efforts that were wholeheartedly supported by the FDA, the World Health Organization, the USDA, and the National Institutes of Health. The solution to the problem of heart disease appeared to be obvious: Replace saturated fats with unsaturated fats. Americans were told to eat margarine instead of butter. Unfortunately, margarine contained a type of fat (trans fats) that was far more dangerous than anyone could have possibly imagined.

To understand what trans fats are, let's go back to our description of an unsaturated fat. In the diagram below, look at the carbon atoms shown in bold. The hydrogen atoms connected to those two carbon atoms are both on the *same* side. This is called being in the "cis configuration." *Cis,* in Latin, means "on this side of." When both hydrogen atoms are on the same side, they repel each other, causing a bend in the molecule. This bend makes it harder to stack one molecule on top of another. Molecules that don't stack well are hard to crystallize or, said another way, they are hard to make into a solid. As a result, cis unsaturated fats are invariably liquid oils, like canola and sunflower oils.

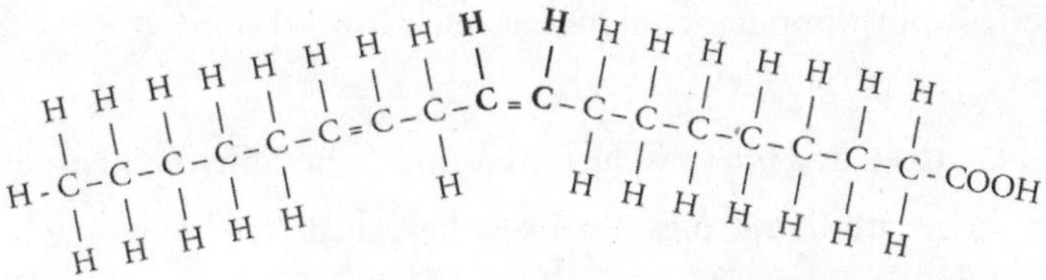

```
                              H H  H H H H H H H H
  H H H H H H H H H           | |  | | | | | | | |
  | | | | | | | | |           C=C-C-C-C-C-C-C-C-COOH
H-C-C-C-C-C-C=C-C-C       ...
  | |  | | |    |              | | | | | | |
  H H  H H H    H              H H H H H H H
```

Cis-Unsaturated Fat

Sometimes, as shown in the example on page 52, the hydrogen atoms of an unsaturated fat are on the *opposite* side. Now the hydrogen atoms are said to be in a *trans* configuration. *Trans,* in Latin, means "on the other side." When hydrogen atoms are on the opposite side, the molecule is straight. Now it's much easier to stack one molecule on top of another. Molecules that stack neatly and tightly are easy to crystallize, converting a liquid into a solid. That's why common vegetable shortenings, even though they are made of vegetable oils, stay solid in a can on your kitchen shelf.

For the most part, large quantities of trans fats are *not* found in nature. They are created when hydrogen atoms are purposely added to unsaturated vegetable oils, a process called hydrogenation. The end product is typically referred to as a "partially hydrogenated vegetable

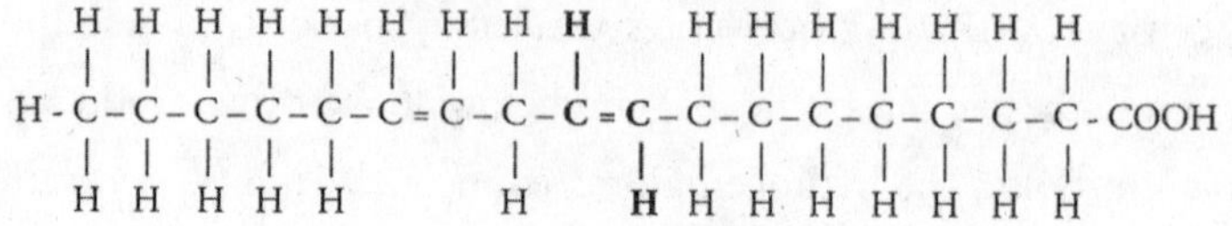

```
   H H H H H H H H H   H H H H H H H
   | | | | | | | | |   | | | | | | |
 H-C-C-C-C-C-C=C-C-C=C-C-C-C-C-C-C-C-COOH
   | | | | |     |   | | | | | | | |
   H H H H H     H   H H H H H H H H
```

Trans-Unsaturated Fat (or Trans Fat)

oil." The designation *partially* means that the product isn't completely saturated or, said another way, is still unsaturated. It also means that the product is loaded with trans fats.

Americans first became aware of unsaturated fats containing large quantities of trans fats in the 1980s. But in truth, these products were actually born more than a hundred years earlier in the form of one of America's most popular cooking products.

On February 27, 1901, Wilhelm Normann became the first person to hydrogenate liquid oils, a process he called "fat hardening." On August 14, 1902, Normann was awarded German patent #141,029. Trans fats were born. One year later, after Normann was granted a patent in England, Joseph Crosfield & Sons built a large-scale manufacturing plant in Warrington, England. By 1909, Crosfield was producing 6.6 million pounds of partially hydrogenated vegetable oils every year. Five years later, more than 20 plants worldwide were hydrogenating vegetable oils into a solid state—all loaded with trans fats.

The same year that Joseph Crosfield & Sons started mass-producing solid oils, Procter & Gamble acquired the U.S. rights to Normann's patent, originally planning to use it to make soaps and candles. Soon Procter & Gamble scientists figured out how to use Normann's method to convert cottonseed oil from a liquid to a solid. When he realized that his company had created a cooking product like no other, William Procter walked into the office of a man who had been selling cooking oils for most of his life, tossed a hard white block onto his desk, and said, "There is some cottonseed oil." They called it Crisco, a contraction of *Crys*tallized *c*ottonseed *o*il.

For many reasons, the partially hydrogenated vegetable oils containing trans fats in Crisco were superior to every other cooking oil or shortening ever invented: (1) Trans fats are more stable when exposed to oxygen, so they have much longer shelf lives than animal fats like butter; (2) trans fats burn only at extremely high temperatures, so cooking oils don't cause much smoke and don't need to be changed as frequently—a godsend to any employee who works all day over a fryer; (3) trans fats have a neutral flavor, so they don't interfere with the taste of any food; (4) trans fats look so much like butter that they can easily replace it; and (5) trans fats are extraordinarily cheap. Starting in the 1930s, they were made from oils left over from crushing soybeans used to make animal feed. Finally, because of their variations in texture, structure, lubrication, and aeration, semisolid fats like Crisco allowed bakers to make cakes fluffier, cookies crumblier, crackers crispier, pies flakier, chicken crunchier, and croissants more delicate.

Procter & Gamble knew they had a gold mine on their hands. They sold Crisco attached to cookbooks that contained a variety of recipes, all of which required Crisco for baking and frying. They marketed Crisco with phrases like, "It's all vegetable! It's digestible!" and "An Absolutely New Product, a Scientific Discovery Which Will Affect Every Kitchen in America." Also, because Crisco is kosher, they promoted it with this tagline: "The Hebrew Race has been waiting 4,000 years for Crisco." In the 1940s, animal fats like butter accounted for two-thirds of all fat consumption in the United States; by the early 1960s, with the increasing use of partially hydrogenated vegetable oils containing trans fats, that ratio had reversed.

According to Judith Shaw in her book, *Trans Fats,* two events launched the partially hydrogenated vegetable oil (trans fat) industry.

The first was legislation passed by Congress in 1956 to build an interstate highway system. This enabled fast-food restaurants like McDonald's, Burger King, Taco Bell, and Chili's to spread across the country. Because partially hydrogenated vegetable oils had a long shelf life, cookies, french fries, fried chicken, and fried fish could now be transported across the country without preservation. The second piece of legislation that was a boon to trans fats passed on September 6, 1958: the Food Additives Amendment. The amendment, which was intended to protect Americans from potentially dangerous additives, stated, "A food shall be deemed to be adulterated if it bears or contains any poisonous or deleterious substance which may render it injurious to health." Unfortunately, food additives used before 1958 (like partially hydrogenated vegetable oils), didn't require FDA approval. Trans fats had been grandfathered in.

In the 1980s, partially hydrogenated vegetable oils became the single most popular product for all baking and frying. By 2001, hydrogenation became the fourth largest food manufacturing process in the world. Also in 2001, the CDC released its data on the annual incidence of heart disease in the United States: 12.6 million Americans had coronary artery disease; 5.4 million had medical procedures for heart disease; and 500,000 people died from heart attacks and related strokes. The price tag for heart disease was about $300 billion a year.

By pummeling companies that used tropical oils like coconut and palm oil and animal fats like butter—all of which were high in what were believed to be evil saturated fats—CSPI and NHSA had inadvertently caused Americans to use a far more dangerous product: trans fats. Suddenly products like margarine, which contained 25 percent trans fats, became the "healthy alternative." By the early 1990s, tens of thousands of products were made using partially hydrogenated vegetable oils. Because they were cheap, kosher, and promoted as heart-healthy alternatives, they flew off the shelves.

In 1981, a group of Welsh researchers sounded the first alarm, publishing a paper claiming that the trans fats contained in partially hydrogenated vegetable oils were linked to heart disease. Nine years later, in the prestigious *New England Journal of Medicine,* two Dutch researchers published findings supporting the Welsh study. For the first time, Americans were starting to realize that not all unsaturated fats were good for you. In 1993, a study done by Harvard researchers showed that if people replaced just 2 percent of the energy from trans fats with other unsaturated fats, they could decrease their risk of heart disease by 33 percent; another study showed that the same decrease in trans fat intake could lessen the risk of heart disease by 53 percent. The Harvard School of Public Health later estimated that eliminating trans fats from the American diet would prevent 250,000 heart attacks and related deaths every year!

Unlike studies of total fat, total cholesterol, and unsaturated fats—where findings had been contradictory or inconclusive—no researcher has ever published a paper showing that trans fats are anything other than one of the most harmful products ever made. As researchers got better at understanding that not all unsaturated fats were the same, the problem with trans fats became painfully clear.

What about cholesterol? Wasn't cholesterol found in the coronary arteries of people suffering from atherosclerosis? Although it is true that cholesterol, which is an essential component of cells, was found in the fatty streaks that blocked coronary arteries, one particular type of cholesterol was present: low-density lipoprotein (LDL) cholesterol, otherwise known as bad cholesterol. The reason that health advocates warned against products that were high in saturated fats was that saturated fats increase LDL cholesterol. But what these advocates didn't realize at the time was that there are two different types of LDL cholesterol. There's the big, fluffy type, which isn't harmful, and the small, dense type—called very low-density lipoprotein or vLDL cholesterol—which is quite harmful. Saturated fats increase the not-so-bad type of LDL cholesterol but don't increase the very bad vLDL cholesterol.

Another type of cholesterol is actually good for you. Called high-density lipoprotein or HDL cholesterol, it removes vLDL from coronary arteries and transports it to the liver where it can be eliminated from the body. Saturated fats neither increase nor decrease the quantity of HDL cholesterol in the blood.

So, in summary, neither saturated fats nor certain types of cholesterol are necessarily bad for you. Trans fats are a different story. Not only do trans fats dramatically *increase* vLDL, the worst kind of cholesterol, but they also dramatically *decrease* HDL, the helpful cholesterol. For that reason, in 2006, an article in the *New England Journal of Medicine* declared, "On a per calorie basis, trans fats appear to increase the risk of coronary heart disease more than any other macronutrient."

ALTHOUGH THE FOOD AND ADDITIVES AMENDMENT had stated that additives used before 1958 did not require FDA approval, one clause in the bill did allow the FDA to act: "Foods must be examined in the light of current scientific information if their use is to be continued." Health activists first petitioned the FDA to limit the use of trans fats in 1994. In 1999, five years later, the FDA finally announced that it would devise a plan to limit the consumption of trans fats. Three years passed. Nothing happened. On July 10, 2002, the Institute of Medicine (IOM) made a statement designed to shock the FDA into action. The IOM reported that no amount of trans fats was safe, recommending an "upper intake level of zero." At the time that the IOM drew a line in the sand, 95 percent of cookies, 80 percent of frozen breakfast foods, 75 percent of snacks and chips, 70 percent of cake mixes, and 50 percent of cereals contained trans fats.

Public advocacy groups eventually regretted their role in inadvertently promoting unsaturated fats containing trans fats. In 2004, the executive director of CSPI said, "Twenty years ago, scientists, including me, thought trans fats were innocuous. Since then, we've learned otherwise." A year later, Walter Willett, a professor of medicine at Harvard Medical School and chair of the department of nutrition at Harvard School of Public Health, told the *New York Times,* "A lot of people had made their careers telling people to eat margarine instead of butter. When I was a physician in the 1980s, that's what I was telling people to do and unfortunately we were often sending them to their graves prematurely."

WHEN HEALTH ADVOCATES THOUGHT cholesterol or total fat or saturated fats increased the risk of heart disease, they simply launched public relations campaigns to inform consumers. Trans fats, on the other hand, were so clearly dangerous that their presence in foods launched government efforts to ban them. It started in Europe.

On January 1, 2004, Denmark introduced legislation to restrict trans fats to no more than 2 percent of the total fat in any food. Consumption of trans fats fell from 4.5 grams a day per person in 1975 to 2.2 grams in 1993 to 1.5 grams in 1995 to almost 0 grams by 2005. By 2010, the incidence of heart disease and related deaths in Denmark had dropped 60 percent.

On January 1, 2006—12 years after it had first been petitioned to act—the FDA finally announced its plan, which required manufacturers of packaged foods to list the quantity of trans fats on every nutrition label. By the end of the year, 84 percent of Americans had heard of trans fats and at least half could correctly identify their health risks. Kentucky Fried Chicken voluntarily eliminated trans fats followed by Applebee's, Arby's, Taco Bell, and Starbucks. Some of the nation's largest food suppliers, like Kraft, Sodexo, and Frito-Lay, which makes Doritos, Tostitos, and Cheetos, also eliminated their use of trans fats. By 2008, the amount of trans fats in prepared foods had decreased by half. By 2012, trans fats had been eliminated from an estimated 10,000 products and had been banned from restaurants in at least 13 U.S. jurisdictions. New York City, for example, asked 20,000 restaurants and 14,000 food suppliers to eliminate their use of partially hydrogenated vegetable oils containing trans fats.

There is, however, one unfortunate loophole. If products contain less than 0.5 gram of trans fats, the FDA allows manufacturers to claim 0 grams of trans fats on the nutrition label. Because many products contain slightly less than 0.5 gram of trans fats, it's still possible to consume more than the 2-gram limit of trans fats a day set by the American Heart Association. For example, crème-filled sponge cakes contain 0.46 gram of trans fats but are listed as having 0 grams on the label. And microwave popcorn, which contains 0.25 gram of trans fats, also is listed as having 0 grams. Trans fats are also still contained in some brands of margarines and coffee creamers. And they're still contained in Berger cookies. The key to avoiding the problem of hidden trans fats is to look for the phrase "partially hydrogenated vegetable oil" on the nutrition label.

EVERY FEW YEARS the Society of German Chemists gives out its Wilhelm Normann Award for outstanding contributions to fat research and fat science. Ironic, given that Normann's process for converting unsaturated fats to trans fats has probably caused more disease and death than any other man-made chemical reaction in history.

So, what's the take-home lesson? Could any of this have been avoided? Again, as was the case with painkillers, **it's all about the data**. In the late 1970s, when the McGovern committee stated that total fat intake should be less than 30 percent of total calories, data weren't available to make such a strong recommendation. Similarly, when recommendations about which type of fat should be preferred were being made, studies were conflicting. Although several studies showed that saturated fats might increase the rate of heart disease, one Welsh study published at the same time showed exactly the opposite: Unsaturated fats increased the risk of heart disease, dramatically. This conflict should have at least given us pause. But it didn't. Ill-founded promises had been let out of the box, and American tables proudly served margarine as the "heart-healthy" alternative to butter when it was exactly the opposite.

# Chapter 3

# BLOOD FROM AIR

*"For earth is full of evils. And the sea is full."*

—Hesiod, *The Works and Days*

We're not that complicated.

Although we come in different shapes and sizes, heights and weights, and backgrounds and temperaments, and although we have different genes that make different proteins and different enzymes, we all boil down to four essential elements: hydrogen, oxygen, carbon, and nitrogen. If any one of these elements becomes unavailable, our time on Earth will end.

Three of the four elements are easily obtained.

Hydrogen comes from the water we drink, which consists of two hydrogen atoms and one oxygen atom ($H_2O$). Oxygen, not surprisingly, comes from the air we breathe ($O_2$). (Only fish, through their gills, can extract oxygen from water.) Carbon also comes from the air. Green plants, in the presence of sunlight, take carbon dioxide ($CO_2$) from the air and capture it in the form of complex sugars that contain carbon (this is called photosynthesis). We get our carbon from eating plants or from eating animals that ate the plants. Either way, because air and water are abundant, hydrogen, oxygen, and carbon are also abundant.

The weakest link in the cycle of life is nitrogen, which comes only from soil. When farmers grow crops like corn, wheat, barley, potatoes, or rice, they deplete nitrogen from the soil. If they don't replace it, the soil won't be rich enough to grow more crops. Farmers replenish nitrogen in three ways. They use natural fertilizers made from decaying plants or animal manure. They rotate their crops with legumes like chickpeas, alfalfa, peas, soybeans, or clover, which harbor bacteria in their roots that take nitrogen from the air and convert it into a usable form in the soil—a process called "nitrogen fixation." Or they wait for thunderstorms; lightning, as it turns out, can also fix nitrogen from the air.

If every farmer in every country on every continent in the world used every inch of fertile land, sprinkled their fields with natural fertilizers, meticulously rotated their crops, and convinced everyone to eat a vegetarian diet, they could feed about four billion people. But, as of 2016, more than seven billion people roamed the Earth. And although pockets of people are starving, the problem isn't that there isn't enough food. There's plenty of food. The problem is that we don't do a good enough job of distributing it to those who need it.

So how are farmers able to do this? How are they able to feed so many people? The answer lies in an event that occurred on July 2, 1909. Because of this singular moment, 50 percent of the nitrogen in our bodies comes from natural sources and 50 percent comes from the work of one man—a man who at once saved our lives and sowed the seeds of our destruction.

FRITZ HABER WAS BORN ON December 9, 1868, in Breslau, Germany. His parents, Siegfried and Paula, were first cousins, marrying despite their family's objections. Tragedy soon followed.

Three weeks after Fritz was born, on New Year's Eve, Paula died from complications of the birth. Siegfried never recovered. Sinking deep into a depression, he buried himself in his work, ignoring his son. As a consequence, Fritz was raised by his aunts, a grandmother, and a housekeeper. Seven years after Paula died, Siegfried remarried, had three daughters in five years, and became a loving, attentive father—to his daughters. He continued, however, to ignore his son, whose presence constantly reminded him of his first wife's death. Fritz spent much of his young life trying to win his father's approval, without success.

One event underlined their broken relationship. After Fritz graduated from high school, he celebrated late into the night at a local pub. Breakfasts at Siegfried Haber's house, however, began at 7:15 a.m. sharp, no excuses, no exceptions. When Siegfried saw that Fritz was still sleeping, he paraded his daughters into his son's bedroom. "Look well!" he warned them. "This is how the life of a drunkard begins!" Forty years later—still unable to reconcile his father's distance and disappointment—Fritz Haber wept while telling this story to a friend.

Failing to win the love of his father, Fritz sought the love of his fatherland, which, despite his enormous accomplishments, would later reject him in the cruelest manner possible.

When he was 19, Fritz entered Heidelberg University. There, under the mentorship of Robert Bunsen, he fell in love with chemistry, studying light emissions from the newly invented Bunsen burner. Unlike his peers, Haber didn't aspire to the life of an academic; he wanted to do something practical, something that made a difference, something that revolutionized an industry. So he left the university and worked in a distillery in Budapest, a fertilizer factory near Auschwitz, and a textile company near Breslau.

When he was 22, Haber returned to Berlin to attend the Charlottenburg Institute of Technology to work with Carl Liebermann, the first scientist to synthesize alizarin, a popular red dye. Fritz saw synthetic dyes as the future, a perfect marriage of his love of chemistry with his insatiable need for his father's approval. Siegfried Haber bought and sold natural dyes; his son fully expected that he would lead his father's company out of the dark ages of natural dyes and into the bright new era of synthetic dyes.

But Fritz didn't distinguish himself as a businessman. In 1892, when a cholera epidemic swept through the port in Hamburg, Germany, Fritz convinced his father to buy all available stores of chloride of lime, the only known disinfectant. When the epidemic quickly subsided, the Habers were stuck with a product of little value. Siegfried called his son a fool and fired him. "Go to a university!" he shouted. "You don't belong in business!"

At the age of 26, Fritz Haber left the dye business to attend the University of Karlsruhe. There, on the Rhine River just south of Heidelberg, he did something that most chemists at the time thought was impossible. For this single discovery, Fritz Haber would win the Nobel Prize. But when he went to Stockholm to receive it, several other Nobel Prize winners boycotted the event, unable to reconcile the atrocities he had committed.

In the fall of 1898, in a music hall in Bristol, England, Sir William Crookes got up to speak. Crookes was the president of the British Academy of Sciences. A chemist and physicist, he had discovered a new element (thallium) and invented a cathode ray tube that would later be used in televisions and computers. The year before his lecture, the Queen of England had knighted him. William Crookes had spent most of his career being right. When he got up to speak, people listened.

Everyone in attendance assumed that Crookes would do what all former presidents of the academy had done: bore them with a list of accomplishments by British scientists. But Crookes went off script with a speech that would later be called one of the best of the century. "England and all nations stand in deadly peril," he began. Crookes explained how advances in science and medicine had created a dilemma. People were living longer. As a result, there were more mouths to feed. Given that all the great plains on Earth were already being farmed, that each acre could feed only about ten people, and that cities were becoming increasingly more populated, it was only a matter of time before there wouldn't be enough food. People of "the civilized nations," said Crookes, were on the verge of starving to death.

Crookes predicted that the dying would begin sometime in the 1930s. First it would be thousands, then hundreds of thousands, then millions. Although scientists argued about *when* this would start, no one argued *that* it would start. The population was growing faster than the world's capacity to feed it. The solution, said Crookes, was in the production of synthetic, nitrogen-containing fertilizer. Scientists needed to find a way to fix nitrogen from the air and convert it into a form that was usable in soil. Nitrogen fixation by legumes and lightning wasn't going to be enough. "The fixation of nitrogen is vital to the progress of civilized humanity," he said. "It is the chemist who must come to our rescue. Through the laboratory, starvation can be turned into plenty."

William Crookes had issued his challenge. Fixing nitrogen in the laboratory to make synthetic fertilizer was now the holy grail of chemistry. But it wasn't going to be easy.

No country knew how important it was to solve the riddle of synthetic fertilizer more than Germany. At the turn of the 20th century, 58 million people lived in Germany, most in densely populated urban centers. German farmers did what they could with what they had, meticulously recycling decaying plants and animal manure. But it wasn't nearly enough. Germany had to import another source of natural fertilizer; without it, the country wouldn't have survived. To get it, German sailors had to cross an ocean.

In South America, the Atacama Desert is rich in natural nitrogen in the form of nitrates. (Nitrates, which are a combination of one nitrogen atom with three oxygen atoms, are an excellent source of natural nitrogen.) Chile owned it. By 1900, Chile was responsible for two-thirds of all the natural fertilizer used on the planet; Germany used a third of that. So great was its need that, by the turn of the century, Germany had imported more than 350,000 tons of nitrates; by 1912, the number rose to 900,000 tons. As a consequence of its reliance on Chilean nitrates, Germany was particularly vulnerable during war—and WWI was just around the corner. Foreign navies could prevent German ships from traveling to Chile, essentially starving German citizens.

HABER'S RISE AT THE University of Karlsruhe was meteoric.

In 1896, two years after he entered the university, Haber wrote a book about physical chemistry that launched his career and enabled his promotion to assistant professor. In 1898, the year of William Crookes's famous speech, Haber wrote a second book that married theoretical and practical chemistry and assured his next promotion. In 1905, he wrote a third book about thermodynamics that catapulted him to a directorship. He was only 37 years old.

Haber dedicated his third book to his wife, Clara, whom he had known since his student days in Breslau and whom he had married four years earlier. Together they had one son, Hermann. Clara was exceptional: Born into a family of chemists, she was the only woman to receive a Ph.D. in chemistry from the University of Breslau and one of the first women in Germany ever to receive a Ph.D. But marriage didn't suit her. A victim of the gender discrimination of her time, she was transformed from a bright, young scientist to a dispirited housewife, forced to watch her husband ignore both her and their son in his quest for fame. "Fritz is so scattered," she said, "if I didn't bring to him his son every once in a while, he wouldn't even know that he was a father."

For Haber, on the other hand, the University of Karlsruhe was bursting with possibilities. The university had an excellent relationship with Badische Anilin & Soda-Fabrik (BASF), a large chemical company just a stone's throw down the Rhine River. If Haber wanted to find practical applications for his work, the association with BASF was perfect. More important, when it came to chemistry and physics, Germany was the place to be. Under Kaiser Wilhelm II, German professors formulated the most brilliant theories, German scientists made the most important discoveries, and German industries had the most advanced facilities. As a result, German academics won more Nobel Prizes than their counterparts in other countries. When Adolf Hitler rose to power, all of this would disappear. But at the time that Fritz Haber entered the University of Karlsruhe, if one were searching for the holy grail of chemistry, Germany would have been the place to find it.

Although Germans had sailed halfway around the world to get their nitrogen, they really didn't have to travel that far to find it. Air is 79 percent nitrogen. Indeed, seven tons of nitrogen circulate above every square yard of Earth's surface. The problem, however, is that nitrogen in the air doesn't exist as a single atom (N). It exists as two atoms coupled together ($N_2$) in a triple bond that is essentially unbreakable—the strongest chemical bond in nature. Although $N_2$ in the air can be used to inflate a million balloons, it cannot be used to grow a single stalk of corn.

$N_2$ is described in chemistry books as odorless, colorless, nonflammable, nonexplosive, nontoxic, and nonreactive. The key word is "nonreactive." $N_2$ is inert, unavailable, dead. If it is to have any biological use—like forming the amino acids, proteins, enzymes, DNA, and RNA of life—then it has to be broken down into two separate atoms. Only then can nitrogen link to hydrogen to form ammonia ($NH_3$) or to oxygen to form nitrates ($NO_3$). Either of these forms can be used in soil to provide nitrogen to crops.

Because $N_2$ isn't commonly broken down by nature, it took an unnatural process to do it—in a sense, an act *against* nature. In 1909, when Fritz Haber became the first person to discover a commercially feasible way to break $N_2$ apart, more than 3,000 articles had been published on the subject. None of them had provided an answer. The planet's stores of nitrogen were slowly depleting. Time was running out.

THE FORMULA is simple:

$$N_2 + 3H_2 \leftrightarrow 2NH_3$$

Reading from left to right, two paired nitrogen atoms combine with three paired hydrogen atoms to form two molecules of ammonia. Ammonia, Haber knew, would be perfect as a synthetic fertilizer.

To force the equation from left to right, Haber used extremely high temperatures and pressures. In 1904, he found that only 0.005 percent of the nitrogen that he had started with on the left side of the equation had ended up as ammonia on the right side—far less than was commercially practical. To help increase yields, Haber tried a variety of catalysts to speed up the reaction: metals like nickel and manganese that provided platforms on which nitrogen and hydrogen atoms were more readily exchanged. But nothing worked. Haber concluded that creating ammonia from nitrogen in the air was impractical. So he gave up. But not before publishing his findings in a leading chemistry journal.

Haber's publication drew the attention of Walther Nernst, the first professor of physical chemistry at the University of Göttingen and a giant in the field. Nernst had spent much of his academic life working on a theory of heat that would later be known as the third law of thermodynamics, an accomplishment for which he won the Nobel Prize. Nernst was upset with Haber's publication because it had contradicted his own theories. So he asked one of his assistants to repeat the experiment, finding yields that were even lower than Haber had published. Angry, Nernst immediately wrote Haber a letter to express his concern.

Haber took Nernst's criticisms to heart and repeated the experiments, finding that Nernst was right. Both men had now done the same experiments and reached the same conclusion. Synthesizing ammonia from nitrogen in the air was far too inefficient. But Walther Nernst wouldn't let things go. Considering Haber's first publication an affront, he was intent on humiliating him. In May 1907, at an international meeting of the Bunsen Society, Nernst called out Fritz Haber. "I would like to suggest that Professor Haber now employ a method that is certain to produce truly precise values," he said. Later, Nernst called Haber's findings "*stark unrichtigen Zahlen*"—"strongly inaccurate"—and said that fixing nitrogen in the laboratory was a fool's game. Haber was hurt and angry that

Nernst had chosen to criticize him in front of his colleagues. Intent on recovering his reputation, he took to the task of creating ammonia from air with a vengeance.

Following the Bunsen Society meeting, a series of events allowed Fritz Haber to accomplish what appeared to be impossible. First, a young chemist from England named Robert Le Rossignol came to his laboratory. Le Rossignol was a skillful and inventive experimenter, eventually designing a small tabletop apparatus made of quartz and iron capable of withstanding temperatures as high as 1832°F, hot enough to melt copper, and pressures as high as 3,000 pounds per square inch, strong enough to crush a submarine. Second, Haber found a catalyst to speed up the reaction: osmium, a rare metal used as a filament in lightbulbs. Third, Haber found a way to cool down ammonia quickly so that it didn't burn up in high heat. Finally, and most important, Haber's mentor at Karlsruhe, Carl Engler, persuaded BASF to fund Haber's experiments; if they worked, BASF would own the patents and Haber would have a commercial partner.

Haber and Le Rossignol tinkered with the fittings and tried different temperatures and pressures. Finally, in March 1909, they had a glimpse of success. Haber was ecstatic. "Come down, you have to see how the liquid ammonia is running out!" he shouted to a colleague, who remembered, "I can still see it. There was about a cubic centimeter of ammonia. It was fantastic." It wasn't much—about a fifth of a teaspoon—but it was a start. Within a few months, Haber and Le Rossignol's apparatus was producing ammonia round the clock.

On July 2, 1909, BASF sent two representatives to visit Fritz Haber's laboratory: Carl Bosch, chief research engineer, and Alwin Mittasch, a chemist and catalyst expert. Unfortunately, one of

Haber's assistants had been too forceful in tightening a seal, which started to leak during the demonstration. Bosch waited while they tried to fix it. Hours passed. Bosch started to look at his watch; then he left. Mittasch stayed behind. After the apparatus was fixed and the reaction restarted, the machine functioned perfectly, making three ounces of ammonia during the five-hour demonstration. The output was logarithmically better than before. Instead of yields of 0.005 percent, they were now around 8 percent. Mittasch was convinced these higher yields were commercially feasible. He drove back to BASF to tell Bosch the good news.

A few ounces of ammonia produced by a few men in a minor university in Karlsruhe, Germany, would soon change the world. And it would make Fritz Haber a very rich man. But there would be a price to pay.

SCALING UP WASN'T going to be easy. BASF now had to convert a tabletop apparatus into an industry. The man who did it was Carl Bosch.

Bosch was only 35 years old when he took charge of the project. The son of a gas and plumbing supplier in Cologne, Bosch had had the run of his father's workshop. As a child, he removed the finish from his parents' bed because he wanted to see the wood underneath. Then, piece by piece, he took apart his mother's sewing machine to see how it worked. As a young man, he frequently visited his father's factory, learning about soldering, pipe fitting, machining, woodworking, and metallurgy—all skills that would come in handy years later.

Ten months after Haber's demonstration, Carl Bosch built a small prototypic unit in Ludwigshafen, a village not far from Karlsruhe. The plant officially opened on May 18, 1910. Haber's two-foot-high tabletop apparatus had become a 26-foot-high megamachine. Within two months, the unit had produced more than 2,000 pounds of ammonia. By the beginning of January 1911, it was producing more than 8,000 pounds a day. Bosch then moved the operation a little farther down the Rhine to Oppau.

The plant at Oppau, which officially opened in September 1913, was the first of its kind. Costing $100 million, Oppau housed more than 10,000 workers, included a shipping system with its own railroad, boasted miles of piping and tubing, and centered on locomotive-size compressors that could withstand levels of heat and pressure never before seen in this industry. The research laboratory, which was five stories high, contained 250 chemists and a thousand assistants. Haber and Le Rossignol's experiments were now conducted at an industrial level. BASF chemists tested 4,000 different catalysts in 20,000 separate experiments. By the end of the year, the plant at Oppau, which operated 24 hours a day, was producing 60,000 tons of ammonia a year! As far as Germany was concerned, Chilean nitrates were now obsolete.

For his efforts, Carl Bosch became the first person ever to win a Nobel Prize for commercializing a technology he didn't invent. The process, however, was not without its tragedies. In 1921, an explosion at Oppau killed more than 500 people.

For his remarkable discovery, Fritz Haber was awarded, feted, and knighted by universities, professional societies, and royalty throughout Europe. The Soviet Union elected him to its National Academy of Sciences. The American Academy of Arts & Sciences elected him an honorary foreign member.

When the dust settled, Haber knew it was time to move out of Karlsruhe and into Germany's cosmopolitan and intellectual center:

Berlin. The offer was impossible to resist. Haber was made director of the Kaiser Wilhelm Institute for Physical Chemistry and Electrochemistry, given a fabulous salary, allotted 300,000 marks a year for operating expenses, provided a villa for himself and his family, and awarded an endowed chair at the University of Berlin. The Kaiser Wilhelm Institute, which was located in a suburb of Berlin called Dahlem, was the start of something new in Germany—a basic science institute funded solely by the federal government. At its peak, the Kaiser Wilhelm Society for the Advancement of Science operated 38 such research institutes throughout the country and employed more than a thousand scientists and 11 Nobel Prize winners. The brightest star in the center of this research galaxy was Fritz Haber's institute in Dahlem.

OTHER COUNTRIES MIMICKED Bosch's process. By 1963, about 300 ammonia plants were in operation and more than 40 were under construction. Today, about 130 million tons of nitrogen are removed from the air and spread across the earth as fertilizer. More than three billion people alive today—and billions more in the future—owe their existence to Fritz Haber and Carl Bosch. Never before have so many people enjoyed so much food.

Perhaps no one recognized the importance of the Haber-Bosch process more than Norman Borlaug, who won the Nobel Peace Prize in 1970 for launching the green revolution. Borlaug had created new species of wheat and rice. But he knew the real hero of the story. "If the high-yielding dwarf wheat and rice varieties are the catalysts that have ignited the green revolution," he said in his acceptance speech, "chemical fertilizer is the fuel that has powered its forward thrust." During the past hundred years, the Haber-Bosch process has remained essentially unchanged, and ammonia is the most synthesized chemical on Earth.

No country has demonstrated the power of the Haber-Bosch process more than China.

In 1972, President Richard Nixon crossed "the bamboo curtain" and visited China. With him was James Finneran, a senior executive from the M. W. Kellogg Company. At the time, China was the most extensive recycler of human and animal waste in the world. Every ounce of natural nitrogen was put onto every piece of cultivable land. It was simply not possible for China to produce more food than it was producing. But it wasn't enough. Peasants began eating their livestock, wild vegetables, soup made from grass, and bark stripped from trees; in some regions, cannibalism was reported. By 1961, an estimated 30 million Chinese had died from starvation. Those who had survived ate rice and a few vegetables; meat was rare; food was rationed. Worse, the population in China was increasing at a rate of ten million a year.

The M. W. Kellogg Company had constructed the world's most efficient ammonia manufacturing plant. Finneran had accompanied Nixon to China because he wanted to help local industrialists build a plant of their own, assuming they would probably build one. They built 13. Within a few years, Chinese fertilizer production doubled; farmers grew massively greater amounts of crops, enough to feed not only people, but also food animals. Meat became plentiful. By 1989, China was the largest producer and consumer of synthetic fertilizer in the world. And, although China is home to one-fifth of the world's population, malnutrition is no longer a problem. Obesity is the problem. In 1982, 10 percent of China's population was obese (weighing 35 percent or more above ideal body weight). By the early 1990s, that number had climbed to 15 percent. Today, more than 30 percent of people living in Beijing are overweight, and childhood obesity has become a national concern.

In 1924, five years after he had won the Nobel Prize for this discovery, Fritz Haber spoke at the Franklin Institute in Philadelphia, extolling the virtues of science. "The banker and the lawyer, the industrialist and the merchant, despite their leading positions in life, are only administrative officials," he said. "The sovereign is natural science. Its progress determines the measure of prosperity of man; its cultivation is the seed from which the welfare of future generations grow." But scientists don't work in isolation. And science unfettered has a darker side.

The largest nitrogen-producing plant in the United States is located in Donaldsonville, Louisiana. Every day the plant consumes a million dollars worth of natural gas, boils 30,000 tons of water from a local river into steam, and produces 5,000 tons of ammonia (2 million tons a year). Every day these 5,000 tons of ammonia are loaded onto railcars, placed onto barges, floated down the Mississippi River, and sprinkled onto corn and wheat fields across the land. But not all of the nitrogen contained in ammonia ends up in crops. Only about a third of the nitrogen layered onto a cornfield, for instance, ends up in a kernel of corn. The rest washes into streams and leaches into groundwater.

The Gulf of Mexico, located next to the Louisiana ammonia plant, is a perfect example of what can happen when no one is watching. Every year about 1.5 million tons of nitrogen are dumped into the Gulf. This excess nitrogen has caused an overgrowth of algae that clouds the water and chokes off oxygen and sunlight to other species, like fish and mollusks. Algal overgrowth has killed streams, lakes, and coastal ecosystems across the Northern Hemisphere. And it's not just the fish that are dying. The birds that eat the fish are dying, too.

The dead zone in the Gulf of Mexico is now the size of New Jersey and is growing. Worse, more than a 150 smaller dead zones have been identified throughout the world. The Baltic Sea north of Germany is one of the most polluted marine ecosystems on the planet; in the 1990s, the Baltic cod industry collapsed. The Thames, Rhine, Meuse, and Elbe Rivers in Europe also contain more than a hundred times the amount of synthetic nitrogen that is considered safe. Similar problems are occurring in the Great Barrier Reef off the coast of Australia, the Mediterranean and Black Seas, and China's two largest rivers: the Huang He and Yangtze. Algae-producing harmful toxins have also been found in the Chesapeake Bay, Long Island Sound, and San Francisco Bay. In fairness, all this excess nitrogen doesn't come solely from nitrogen fertilizer placed on land. It also comes from manure generated from food animals. But those animals are invariably fed with crops grown with synthetic fertilizer.

Synthetic nitrogen pollution isn't limited to the waters; it has also entered the air and come back to Earth as acid rain, further damaging lakes, streams, and forests as well as the animals that depend on them. These problems will only worsen.

CONTAMINATION OF THE environment with synthetic nitrogen wasn't the only evil that flew out of Pandora's box when Fritz Haber dared to open it. There were a couple more. One involved another process that required the ready availability of nitrogen—a process that explains why Germany stopped making synthetic fertilizer during WWI and moved its production facility from Oppau to Leuna, where it could be better guarded.

When Fritz Haber moved from Karlsruhe to Berlin in 1911, he was able to strike up a friendship with another German scientist: Albert Einstein. Einstein tutored Haber's son in math while Haber helped Einstein through a difficult separation from his wife. "Without Haber," recalled Einstein, "I wouldn't have been able to do it."

Although they were friends, Haber and Einstein couldn't have been more different. Einstein was a liberal, wisecracking, irreverent, bohemian—disgusted by the militarism of his country. Haber was a straitlaced, pro-Kaiser Prussian who believed that German scientists should serve the fatherland whenever asked. On August 4, 1914, when Germany invaded Belgium—a neutral country—in an attempt to outflank France, Haber signed a manifesto defending his country against the international condemnation that quickly followed. Ninety-three German scientists signed the manifesto, including three current and three future Nobel Prize winners. Einstein, on the other hand, was part of a pacifist counterstatement decrying his country's actions. Einstein left the country. Haber enlisted in the army.

German military officials assumed that they would quickly march through France, putting an end to the war. A few months, at most. It didn't work out that way. German soldiers were stopped cold at the Marne River near Paris. German military officials now realized that this was going to be a different kind of war. One slugged out in the trenches. And one that would require massive amounts of one of the world's most powerful explosives: ammonium nitrate. (In 1995, Timothy McVeigh used ammonium nitrate purchased from a fertilizer company to blow up a federal building in downtown Oklahoma City, killing 168 people, including many children; injuring 680 others; and destroying or damaging more than 300 buildings within a one-mile radius. All caused by a single, ammonium nitrate-containing truck bomb.)

As the war came to a standstill, the German military became desperate for more explosives. Haber saw an opening. He convinced Carl Bosch that it was possible to convert ammonia into ammonium nitrate using a commercially feasible, one-step process and that the facility at Oppau was the perfect place to do it. Although Bosch initially disagreed, he eventually yielded. By May 1915, the Oppau plant was producing more than 150 tons of ammonium nitrate a day. BASF was no longer just a chemical company; it was an instrument of war. Whereas workers at Oppau had been working round the clock to feed people, now they were working round the clock to kill them. *Brot aus luft,* "bread from air," had become *blut aus luft,* "blood from air." Bosch called the transformation "this dirty little business."

On May 27, 1915, French planes bombed the Oppau factory. In response, another ammonium nitrate–producing facility was built deep within Germany's interior, in a little town near Leipzig called Leuna. On April 27, 1917, the Leuna plant opened. Centered on 13 large smokestacks, the plant, which boasted more than 30,000 workers, was two miles long and a mile wide. It looked like a small city. When the first batches of ammonium nitrate were produced at Leuna, workers scrawled "death to the French" on the canisters. Soon Leuna was producing more than 240,000 tons of ammonium nitrates a year, all of which fed directly into Germany's war machine. Leuna was the largest chemical complex on Earth.

Fritz Haber was in his glory. WWI had now become "the chemist's war," and Haber, as director of the Kaiser Wilhelm Institute, was the chief chemist. He was named *Geheimrat* (privy counselor), a top adviser to the high command and a clear recognition of the importance of science in Kaiser Wilhelm's Germany. He was also made a captain in the German Army, unprecedented for someone who wasn't a soldier. Determined to look the part, Haber shaved his head, commissioned a tailor to make his uniform, and carried himself with a military bearing. Albert Einstein lamented his friend's transformation.

"Haber's picture unfortunately is to be seen everywhere," said Einstein, after a visit to the Kaiser Wilhelm Institute. "It pains me every time I think of it. Unfortunately, I have to accept that this otherwise so splendid man has succumbed to personal vanity." "He wanted to be your best friend and God at the same time," recalled Lise Meitner, a physicist who would later participate in Nobel Prize–winning work on nuclear fission.

On November 9, 1918, Germany surrendered. Although defeated, Germany's war minister, Heinrich Scheüch, appreciated Haber's contributions. "During the long duration of the war you put your broad knowledge and your energy in the service of the fatherland—beyond all measure," he wrote. "You were able to mobilize German chemistry. It was not given to Germany to emerge victorious from this war. That it did not succumb to the supremacy of its enemies after the first few months because of lack of powder, explosives, and other combinations of nitrogen, is in the first instance your achievement. Your splendid success will always live on in history and remain unforgotten." (The Leuna plant, which during WWII was more heavily guarded than Berlin, later fueled Hitler's armies. On May 12, 1944, the U.S. Eighth Air Force sent more than 200 airplanes to bomb Leuna. By the end of the war, 6,000 Allied planes had dropped more than 18,000 tons of explosives on the plant. When the war was over, Albert Speer, the Third Reich's architect, said that if the Allies had focused solely on eliminating the plant at Leuna, WWII could have ended in eight weeks.)

In 1919 when Fritz Haber received the Nobel Prize in chemistry, he wasn't the only German to win it. Both Max Planck, for his work in quantum physics, and Johannes Stark, for his work on the Doppler effect, had also won the Nobel Prize that year. Haber was as proud for his countrymen as he was for himself—proud that German scientists were honored despite the ill will created by WWI. "I think it was a deed of greatness on the part of the Swedish academy to elect three Germans—and only Germans—as prizewinners," he said. "My heartfelt wish is that it may lead to renewed international understanding." It didn't. Two Frenchmen, who had won Nobel Prizes, rejected their prizes in protest; one said that Haber was "morally unfit for the honor." One American, who had won the Nobel Prize five years earlier, also refused to attend—the first such boycott in the history of the event. During the awards ceremony, several other scientists refused to shake Haber's hand. Their disdain wasn't because Haber had figuratively waved the German flag in their faces during his acceptance speech, or because he had unleashed a flood of fixed nitrogen that would choke off estuaries and waterways, or because he had signed a manifesto supporting Germany's aggressive entry into WWI, or because he had fueled the German Army with explosives made from ammonium nitrate. It was because of something else that Fritz Haber had done during the war—another evil that he had allowed to escape into the world.

When it had become clear that German hopes for a quick end to WWI was impossible, that this was going to be a war of attrition, Fritz Haber saw his moment. Not only would he supply the ammonium nitrate that would allow his beloved fatherland to have an almost limitless supply of munitions, but he would also use his knowledge of chemistry to win the war in a different way—one that had never been used before. Haber believed that Germany would win not because its soldiers were braver or its military leaders shrewder, but because its chemists were smarter.

Under Fritz Haber, the Kaiser Wilhelm Institute, which was now surrounded by barbed wire and military guards, had become an integral part of Germany's war machine. The institute's budget, which included the salaries of 1,500 people and 150 scientists, was 50 times larger than during peacetime. In 1916, Haber was named chief of the Chemical Warfare Service. He wanted to find a way to kill the enemy without guns or mortar. To find something that would creep along the ground, seep into the trenches, and kill Allies on the spot. For months he and his team had studied the effects of poison gases on experimental animals (mostly cats), defining the relationship between the concentration of gas and the time of exposure. Haber found that low concentrations of poison gases over longer time intervals could kill just as completely as large concentrations over shorter intervals. His formula for death was later called Haber's constant. By 1918, more than 2,000 scientists in Germany were working on chemical warfare.

Haber wasn't the first to use gas during war. Indeed, the French and Brits had already used tear gas in 1914. But the purpose of tear gas was to temporarily disable the enemy. Haber's goal was to kill them. Haber's gas of choice was chlorine, favored because it was heavier than air—so it could spread down into the trenches—and because it caused almost immediate death, like being suffocated with a poisonous pillow.

When Fritz Haber was performing his experiments, he knew that chemical warfare was a violation of international law. Several years earlier, in 1907, Germany, along with 24 other nations, had signed the Hague Conventions, which forbid countries from "employing poisons and poisonous weapons." Although poison gas was a clear violation of the conventions, Haber didn't care. The goal was to win. And if he violated the rules, so be it. For his actions, Haber would later be branded as a war criminal.

It happened on Thursday, April 22, 1915, at a battleground near the ancient trading city of Ypres, France—the site of one of the bloodiest battles of the war. German soldiers were on one side; French, British, Algerian, and Canadian soldiers on the other. At 5:00 p.m., Fritz Haber opened the valves on 6,000 canisters containing 150 tons of deadly chlorine gas. Standing at his side were three young scientists: Otto Hahn, Gustav Hertz, and James Franck, all of whom would later win Nobel Prizes. Hans Geiger, who would later invent the Geiger counter, was also there. The wind was right. Shortly after the canisters were opened, a yellow-green cloud four miles long rose to the height of a blue whale, traveled southward, and swept toward an unsuspecting battalion of French and Algerian soldiers. Within minutes, birds fell from the sky, leaves shriveled, and thousands of soldiers were choking, gagging, seizing, and turning blue. Those who weren't immediately affected dropped their guns, threw down their packs, and ran. One British soldier recalled the event: "Suddenly down the road from the Yser Canal came a galloping team of horses, the riders goading their mounts in a frenzied way; then another and another, till the road became a seething mass with a pall of dust all over it." The chlorine gas released that day killed 5,000 soldiers and disabled 15,000 others.

Fritz Haber had found a way to escalate the horror of war—a fact not lost on Allied generals. "It is impossible for me to give a real idea of the terror and horror spread among us by this filthy loathsome pestilence," said one Canadian officer. Calling the event "spectacular," Haber knew that his weapon had created not only a technological but also a psychological advantage. "Every new weapon is capable of winning a war," he recalled. "Every war is a war against the soul of the soldier, not the body. New weapons break the morale because they are something new, something that he has not experienced, and, therefore, something that he fears. The artillery did not do much harm to morale, but the smell of gas upset everybody." Not everyone in Germany was applauding. "The higher civilization rises," wrote one German commander, "the viler man becomes." Using the doublespeak that would later become a cynical theme in Hitler's Germany, the attack was called "Operation Disinfection."

After the attack at Ypres, both President Woodrow Wilson and the International Red Cross protested the use of chemical weapons, but to no avail. George Grosz, a prominent German artist, who was 24 years old at the time, protested in his own way. Grosz drew a picture of Christ on the cross with a gas mask and army boots—his attempt to show the bestiality of war, our seemingly limitless capacity for depravity. He was later tried and convicted for blasphemy.

Fritz Haber was unrepentant, declaring that his science belonged to humanity during peacetime but to the fatherland during war. His only regret was that the Germans hadn't taken full advantage of the hole blown into the center of the Allied front. Advancing too slowly, the Germans had allowed Canadian forces to fill in. Haber believed that had the officers been bolder—and more willing to ignore the fact that 200 German soldiers were also poisoned and 12 killed by the gas—the war could have been won that day.

A WEEK AFTER the chlorine gas attack at Ypres, Fritz and Clara Haber hosted a dinner party. When it was over, they got into a terrible fight. Chemical warfare wasn't science, said Clara, it was "a perversion of science." Shy and quiet by nature with a slight lisp, this wasn't an easy conversation for Clara. She rarely questioned her husband's decisions. But Fritz had crossed a line. And Clara couldn't live with it anymore. After Fritz had fallen to sleep, she went into the bedroom, took her husband's handgun, walked outside to the garden, and fired a shot into the air. Convinced that everything was in working order, she aimed the gun at her chest and fired a second time. Her son, Hermann, who was 14 years old, rushed to his mother's side. Clara was bleeding badly but still alive. Hermann screamed for his father, but to no avail. Fritz had taken a sleeping pill and was unarousable. That night, May 2, 1915, Clara Haber died from the self-inflicted gunshot wound. The next day, Fritz Haber left his son and traveled to the Eastern Front, where he was expected. Two years after Clara's suicide, Haber remarried. Thirty years after the suicide, when he was 44 years old and living on Long Island as a patent attorney, Hermann Haber also took his life.

FRITZ HABER NEVER UNDERSTOOD his wife's objections to chemical warfare. And never understood why several of his fellow Nobel Prize winners boycotted his acceptance speech. To Haber, a dead soldier was a dead soldier. It didn't matter *how* they died. It only mattered *that* they died. Poison gas worked to the advantage of technologically advanced societies, so why shouldn't Germany use its assets? "The disapproval that the knight felt for the man with a gun," said Haber, "is repeated by the soldier who shoots bullets [at] a man with chemical weapons." Haber's goal was to turn warfare into a competition among scientists; the winner would make the deadliest poison gases, distribute them most efficiently, and create the best protective gear, including gas masks. "Gas weapons and gas defense turn warfare into chess," he said, dispassionately. Further, using a rationalization that followed the dropping of the atomic bomb in WWII, Haber argued that chemical weapons would save more lives than take. In truth, Fritz Haber was enormously proud of what he'd done. Proud that science had advanced beyond shooting and shelling to something far more devastating. Something that, according to Haber, had turned "soldiers from a sword in the hand of their leader into a heap of helpless people."

Ypres was the first of five chlorine gas attacks. Between April 22 and August 6, 1915, the Germans released 1,200 tons of chlorine gas in five separate attacks on the Allies. BASF's Oppau plant, once a site for manufacturing synthetic fertilizer, now made explosives and poison gas only. By the end of 1915, BASF was making 16,000 tons of chlorine gas a year.

Fritz Haber was the first to use chlorine gas in war. On October 15, 1915, he became the first to use phosgene gas, which, like chlorine, asphyxiated its victims, but required much less gas to do it. At the front in Champagne, between October 15 and 27, 1915, the Germans released 500 tons of phosgene gas, disabling 5,700 Allies and killing 500.

Haber wasn't finished. In 1917, he became the first person to release the most dangerous chemical used in warfare: mustard gas. Unlike chlorine and phosgene gases, which eventually dissipate in the wind, mustard gas hangs around, sticking to soil, clothing, homes, and tools—almost impossible to wash off. Mustard gas caused severe conjunctivitis, making it difficult for soldiers to see; intense inflammation of the skin, mouth, throat, and windpipe, making it difficult to swallow and breathe; widespread blistering similar to second-degree burns; and overwhelming bronchitis and pneumonia, the most common cause of death. Three of every hundred soldiers exposed to mustard gas died. Because mustard gas was the most poisonous of the gases; because it lingered, disabling soldiers long after canisters had been opened; because it had a high death rate; and because soldiers feared mustard gas more than any other chemical weapon, Fritz Haber labeled it "a fabulous success."

Not surprising, mustard gas was the chemical weapon most likely to cause collateral damage. On July 20, 1917, the Germans bombed the western outskirts of Armentières with mustard gas. Thousands of local farmers and townspeople went to their shelters. When they returned and came in contact with brickwork and household objects still laden with the poison, 675 were injured and 86 died.

Between 1914 and 1919, Germany made 87,000 tons of chlorine gas, 24,000 tons of phosgene gas, and 7,700 tons of mustard gas. Although Britain and France would also use chemical weapons, the Germans were the first to enter the race, and by far the most successful. The Germans were also the first to load poison gases into shells and fire them at the enemy. In 1918 about one-third of all German shells contained poison gas. By the end of the war, with Fritz Haber as the ringmaster, more than a million people had been disabled and 26,000 killed by chemical weapons.

World War I officially ended on June 28, 1919, with the signing of the Treaty of Versailles. The following summer, Fritz Haber heard that he was on a list of war criminals and that the Allies had demanded his extradition. In response, he grew a beard, bought a forged passport, fled to Switzerland, obtained citizenship, and settled in St. Moritz. In November 1919, he got word that he had won the Nobel Prize. When the Allies withdrew their extradition request, he returned to his beloved Germany.

Although the Hague Convention of 1907 had been clear about the prohibition of poison gas, the Treaty of Versailles made it even clearer. The Allies wanted to make sure that Germany never used chemical weapons again. Germany was prohibited from using "asphyxiating, poisonous, and other gases and all analogous liquids."

Further, "their manufacture and importation are strictly forbidden." Fritz Haber didn't see it that way. He believed that the treaty had no moral or legal legitimacy. And, as he had done with the Hague Conventions, he ignored it. At the Kaiser Wilhelm Institute, with a framed picture of the chlorine gas attack on Ypres in his office, Haber continued to test chemical weapons on animals. When international weapons inspectors visited his institute, he insisted that he was working on insecticides. And although Haber never imported chemical weapons, as specified by the treaty, he exported them. He helped Spain with its construction of a mustard gas facility. And he helped Russian officials launch a poison gas program at Volga. In 1924, in collaboration with the German Ministry of Defense, he set up chlorine and mustard gas production plants in central Germany, misrepresenting them to foreign inspectors as "oil and refinery plants."

ON JANUARY 30, 1933, Adolf Hitler came into power. Three months later, Hitler introduced his Law for the Restoration of the Professional Civil Services; Jews were no longer allowed to work for the government.

Initially, Fritz Haber didn't think this new law had anything to do with him. He was, after all, a Protestant, having been baptized at Saint Michael's church in Jena when he was 24 years old. But both of Haber's parents were Jewish. So, in the eyes of the Third Reich, Fritz Haber was a Jew. There was, however, an out. At the insistence of Hitler's predecessor, Paul von Hindenburg, everyone who had served their country faithfully in WWI could still be employed by the government, even if they were Jewish.

At the time of Hitler's ascension, about 500,000 Jews lived in Germany, less than one percent of the population. About 10,000 had converted, primarily to make it easier to advance in business and academia. Like most, Haber had converted to make himself "more fully German." Although he'd converted, he did nothing to hide his Jewish ancestry. Both of his wives were Jewish, as were most of his friends. When Haber was asked by the Nazis to fill out a form declaring his ancestry, he wrote "non-Aryan."

On April 21, 1933, Fritz Haber received a phone call from Bernard Rust, head of the Nazi Ministry of Art, Science, and Popular Culture. Rust was clear about what he wanted. Haber had to start firing Jews who worked in his institute. At first, Haber tried to comply. He fired two senior Jewish scientists, but only after he had found them jobs outside of Germany. He couldn't bring himself to fire others, especially younger Jewish scientists, whom he felt needed his protection the most.

Unlike Haber, many German Jewish scientists never converted to Christianity. And all were sickened by Hitler's insistence that they resign their positions. James Franck, who, like Haber, had served his country in WWI, and who, like Haber, later won a Nobel Prize, refused to live in a place where Jews were treated so heinously. So he resigned his post as professor at the University of Göttingen. But not before he wrote a letter to Fritz Haber. "I can't get up in front of my students and act as though all this doesn't matter to me," he wrote. "And I also can't gnaw on the bone that the government tosses to Jewish war veterans. I honor and understand those who want to hold out in their positions, but there also have to be people like me. So don't scold your James Franck, who loves you." (Franck later emigrated to the United States and worked with Robert Oppenheimer on the atomic bomb.)

Franck had assumed that because Haber was the most prominent scientist in the Kaiser Wilhelm Society, he would stay in Germany.

But he was wrong. Fritz Haber had had enough; he needed reassurance that he would continue to be seen as indispensable to the new regime, even though he was Jewish.

THREE WEEKS AFTER Adolf Hitler introduced the law prohibiting Jews from working for the government, Fritz Haber resigned as director of the Kaiser Wilhelm Institute. No one could believe it, including Haber himself, who assumed that Bernard Rust would never accept his resignation—never allow someone as prominent and well respected as him to walk away from his country. "I herewith request that on 1 October 1933 I be allowed to retire," wrote Haber. "According to the directives of the National Civil Service Law of 7 April 1933, I have the right to remain in office, although I am descended from Jewish grandparents and parents. But I do not wish to make use of this dispensation any longer."

When Albert Einstein heard that Haber had resigned, he immediately sent him a letter. "I can imagine your inner conflicts," he wrote. "It is somewhat like having to abandon a theory on which you have worked for your whole life. It's not the same for me because I never believed in it in the least." The theory, presumably, was that the German high command would honor faithfulness, blind devotion, and service; that they would be decent; that they would care about what Fritz Haber could do for them and not what religion he had been born into. But they didn't. Haber was a Jew and that was all the Nazis could see. They didn't care about scholarship or academia. On the contrary, German scientists, scholars, and intellectuals were more a threat than a matter of civic pride. When Rust accepted his resignation, Haber was stunned. "I am bitter as never before," he wrote.

Several scientists tried to intervene on Haber's behalf.

First, friends of Haber asked Rust to reconsider. But Rust was unwilling to relent. "I'm finished with the Jew, Haber," he said.

Then, Max Planck, who had won the Nobel Prize in physics and, like Haber, was enormously respected in scientific circles, met with Adolf Hitler. Planck argued that Haber's resignation was bad for German science; that although Jews made up only one percent of the population, one-third of Germany's Nobel Prize–winning scientists were Jewish. It was "self-mutilation," argued Planck. Hitler would have none of it. As the meeting progressed, Hitler talked faster and louder, pounding his fist on his knees, screaming, possessed. Planck, who was 75 years old, left the room without looking back. Shaken, it took him days to recover.

Finally, Carl Bosch, who was now Germany's foremost chemical industrialist, also met with Hitler, arguing that persecuting Jewish scientists was bad for German business. But, as with Planck, Hitler seemed to be in a trance, a dream. Referring to what he believed would be his hundred-year Reich, Hitler screamed, "You don't understand these matters! If Jews are so important to physics and chemistry, then we'll just have to work one hundred years without physics and chemistry." Bosch, who would become openly critical of Hitler's policies, was later relieved of his executive duties. He died in 1940 from depression and alcoholism.

On August 3, 1933, Fritz Haber left Germany. Looking for work, he traveled from hotel to hotel in Spain, Holland, France, England, and Switzerland. Most of Haber's colleagues, however, couldn't forget his notorious past as an unabashed promoter of chemical warfare. Ernest Rutherford—a British scientist and the father of nuclear physics—refused to meet with him. After a few months, Haber was offered a meaningless position at the University of Cambridge. Still, he considered it—anything to rid himself of the stain of his native land. "My most important goal in life is that I not die as a German citizen," he said.

While in Switzerland, Haber met Chaim Weizmann, a Russian-born, Jewish scientist and leading proponent for a Jewish homeland in Palestine. Weizmann remembered his first meeting with Haber. "He was a broken man," said Weizmann, "moving about in a moral vacuum. I made a feeble attempt to comfort him, but the truth is that I could scarcely look into his eyes. I was ashamed for myself, ashamed for this cruel world, and ashamed for the error in which he had lived and worked throughout his life." But Weizmann saw in Haber a man who had now finally embraced his Jewish heritage. So he offered him a job at the Daniel Sieff Institute (now the Weizmann Institute) in Rehovot, just outside of Tel Aviv. "You will work in peace and honor," said Weizmann. "It will be a return home for you."

Haber was humbled and overwhelmed by Weizmann's offer. "Dr. Weizmann, I was one of the mightiest men in Germany," he said. "I was more than a great army commander, more than a captain of industry. I was the founder of industries; my work was essential for the economic and military expansion of Germany. All doors were open for me. But the position I occupied then, glamorous as it may have seemed, was nothing compared with yours. You are not creating out of plenty—you are creating out of nothing, in a land that lacks everything; you are trying to restore a derelict people to a sense of dignity. And you are, I think, succeeding. At the end of my life I find myself bankrupt. When I am gone and forgotten, your work will stand, a shining moment in the long history of our people." Haber's life had come full circle. After rejecting his Jewish heritage, he now fully embraced it. To paraphrase T. S. Eliot, he had arrived where he had started and knew the place for the first time.

Perhaps no one was more surprised by Haber's transformation than his friend of 20 years, Albert Einstein. Haber wrote to Einstein after his meeting with Weizmann: "In my whole life, I have never been as Jewish as I am now!" Einstein wrote back: "I was very happy to receive such a detailed and long letter from you, and was especially happy that your earlier love for the blonde beast has cooled a little. Who would have thought that my dear Haber would approach me as the advocate of the Jewish, even the Palestinian, matter! I hope you will not return to Germany. It is no true business to work for an intelligentsia consisting of men who prostrate themselves on their bellies before common criminals and even sympathize to a certain degree with these criminals." Einstein ended his letter with a wish: "I hope to meet you soon under a milder sky." But Albert Einstein and Fritz Haber would never meet again.

While traveling to Zermatt, Haber was taken off the train in the small Swiss town of Brig with chest pains. His sister, Elyse, immediately came to look after him. Suffering from severe heart disease, Haber was treated with the medications of his time: nitroglycerin and bloodletting. On January 29, 1934, almost one year to the day after Adolf Hitler rose to power, Fritz Haber died. He was buried in Basel, Switzerland. Haber requested that his tombstone be engraved with the sentence, "In war and peace, as long as it was granted him, a servant of his homeland." Fritz's son, Hermann, ashamed of his country's atrocities, couldn't bring himself to honor that wish. One request was honored. Even though he had been married a second time, Haber asked that he be buried next to his first wife, Clara. So Clara's body was exhumed from Dahlem, Germany, and transported to Basel, Switzerland, where both are now buried under a tombstone that bears only their names and dates. Perhaps the most fitting epitaph was written by Albert Einstein who, when he learned of his friend's death, wrote a letter to Hermann lamenting that Haber had suffered "the tragedy of the German Jew; the tragedy of unrequited love."

### One final irony.

On February 15, 1917, Fritz Haber and several German industrialists met to discuss the best methods for pest control, primarily for agriculture. The meeting spawned the Technical Committee for Pest Control; Haber was the chairman. The best way to control pests, all agreed, was with hydrogen cyanide (HCN). The question was how to administer it.

At the time of the meeting, body lice, which carried deadly typhus bacteria, plagued German soldiers at the front. And moths were a huge problem in flour mills. Under the supervision of Fritz Haber, German scientists developed methods to administer HCN. First, they simply released the chemical from steel canisters. Then they developed a vat method, where sodium or calcium cyanide was added to vats of sulfuric acid, releasing HCN gas. Finally, they developed a method where HCN pellets were exposed to hot air, releasing the gas. (The final product was called Zyklon.) Although many countries used HCN as a pesticide, no country used it more efficiently or more universally than Germany. The Germans effectively fumigated granaries, barracks, trains, warships, and entire buildings, which were emptied, sealed, and pumped with Zyklon.

One problem with HCN gas was that it was odorless and colorless. As a consequence, some people unknowingly exposed to HCN gas died. To prevent accidental exposure, Haber and his team added cyanogen chloride, a benign chemical that gave the gas a foul odor.

The foul-smelling preparation of Zyklon was called Zyklon A. In 1920, the inventors of Zyklon A moved to another institute, although Haber still retained an association. There, they developed Zyklon B, later used by the Nazis to kill more than a million Jews in concentration camps, primarily Auschwitz and Treblinka. (The Nazis removed the cyanogen chloride so that those being gassed wouldn't know what was coming.) Several of Fritz Haber's relatives died there, including the daughter of his half sister, Frieda (Hilde Glucksmann), her husband, and their two children. Frieda was the daughter of Siegfried Haber's second wife.

Although Haber had supervised the manufacture of Zyklon—a chemical that would kill millions of his fellow Jews—he could never have conceived of its eventual use. Indeed, at the beginning of Adolf Hitler's rise to power, Fritz Haber realized that he had inadvertently provided the Nazis with the munitions and chemicals necessary for their reign of terror. "I have put fire in the hands of small children," he lamented.

IN THE DEUTSCHES MUSEUM in Munich, separated from onlookers by a small barrier, stands the tabletop device Fritz Haber and Robert Le Rossignol built to fix nitrogen from the air. Onlookers occasionally stop, stare for a few seconds, and walk past, thinking little of this machine that launched the worldwide manufacture of synthetic fertilizer, a process that has given so many people their lives and—due to ongoing contamination of the environment with excess nitrogen—a process that has probably started the clock on their eventual destruction.

FRITZ HABER HAS ALLOWED three billion more people to live on the face of the Earth than would ever have been possible. His accomplishment is nothing short of phenomenal. The price tag for Haber's invention, however, has been the gradual death of streams, lakes, waterways, and oceans. The lesson here is this: **Everything has a price; the only question is how big**. Most people would be surprised to know that even the most dramatic, lifesaving, medical and scientific breakthroughs like vaccines, antibiotics, and sanitation programs have unintended and occasionally tragic consequences. We'll talk about this in the last chapter.

# Chapter 4

# AMERICA'S MASTER RACE

*"Every good tree bringeth forth good fruit; but a corrupt tree bringeth forth evil fruit."*

—Matthew 7:17

On June 16, 2015, real estate mogul and *Apprentice* star Donald Trump launched his bid for the Republican nomination for president of the United States. He did it by attacking Mexican immigrants. If Americans wanted to know how their country could become great again—how they could rid themselves of the social, political, and financial woes of the recent decade—all they had to do was look south of the border. There they would find their bogeyman. "When Mexico sends its people, they're not sending the best," said Trump. "They're sending you people that have lots of problems, and they're bringing their problems with [them]. They're bringing drugs. They're bringing crime. They're rapists."

The facts tell a very different story: (1) First-generation Mexican immigrants actually commit fewer crimes than native-born Americans; (2) as rates of immigration have increased, rates of crime have decreased; and (3) the percentage of illegal immigrants in prison is actually less than that in the general population. The reasons are obvious. Because they risk deportation, undocumented immigrants have a strong desire to stay out of trouble. "Immigrants in general—unauthorized immigrants in particular—are a self-selected group who generally come to the U.S. to work," said Marc Rosenblum, deputy director of the U.S. Immigration Policy Program. "And once they're here, most of them want to keep their nose down and do their business; they're sensitive to the fact that they're vulnerable."

Trump had found a way to galvanize the American public. When he made Mexican immigrants the cornerstone of his campaign, his favorability among Republican voters leaped from 16 percent to 57 percent, a spike greater than any of his challengers. Other candidates vying for the Republication nomination began echoing his ideas, calling for increased security and fencing. Ted Cruz, a senator from Texas and a Hispanic American himself (though of Cuban descent, not Mexican), noted in his immigration plan that "the unsecured border with Mexico invites illegal immigrants, criminals, and terrorists to tread on American soil."

As the Republican primary went on, the attacks were not limited to Mexican immigrants. In response to a terrorist attack in San Bernardino, California, in December 2015, Trump and other conservative politicians called for "a total and complete shutdown of Muslims entering the United States." Not only were illegal immigrants from Mexico "murderers" and "rapists," now, apparently, legal immigrants of the Muslim faith were potential terrorists. The ban, which would apply to about a billion Muslims, presumably would include those visiting family members, academics, and parents seeking specialized medical care for their children. "Anyone who thinks [Trump's] comments will hurt him don't know the temperature of the American ppl," tweeted radio host Laura Ingraham. At the time, 59 percent of

Republicans supported the ban, but only 36 percent of the general American public viewed it positively. But just a few months later in March 2016, 51 percent of Americans expressed support for a ban on Muslims entering the country.

These politicians had tapped into a common thread in American history—fear of immigrants. Whether the rejection of mostly Jewish immigrants from Eastern Europe in the 1930s and 1940s or, when compared with Canada and other countries, the woeful rate of acceptance of Syrian immigrants today, America has often been slow to open its doors.

What most Americans had failed to realize, however, was that these successful appeals to our worst prejudices had their roots in a scientific treatise published a century ago. The author was a New York City conservationist named Madison Grant.

It started with pea plants.

In 1866, a dyspeptic, curmudgeonly, Augustine monk working in Brno, Moravia, published a scientific paper in the *Proceedings of the Natural History Society of Brünn*. No one noticed. The monk's name was Gregor Mendel, and the subject was peas. Mendel wondered what would happen if he crossed a tall pea plant with a short one. Would the offspring be tall, short, or something in between? And what about pea plants with wrinkled pods or smooth pods or with green leaves or yellow leaves? What he found surprised him; there was no in between. Progeny were either tall or short; pods were wrinkled or smooth; and flowers were green or yellow. Traits never blended; rather, certain traits seemed to dominate. Tall was dominant over short, wrinkled over smooth, and green over yellow. Mendel didn't base his conclusions on a handful of plants or a few years of study; he had worked for more than a decade and performed thousands of cross-fertilizations. At the end of his paper, Mendel proposed that his peas were inheriting one "factor" from each parent. Today we call these "factors" genes.

Despite popular belief, Gregor Mendel didn't discover heredity. At the time of his publication, people knew that cows could be bred to produce more milk or chickens to produce more eggs or horses to produce more wins at the racetrack. What they didn't know was why. Because of Gregor Mendel, people could now predict whether an animal would express a particular physical trait. Mendel had moved animal breeding into the realm of computational biology.

A few years after Mendel published his paper, a British scientist named Francis Galton, who was a half cousin of Charles Darwin, made the leap from peas to people and from physical traits to something far more meaningful. If we could breed better animals, reasoned Galton, couldn't we breed better people, too? Wouldn't traits like intelligence, loyalty, bravery, and honesty also be inherited? And wouldn't selecting for better people make for a better world? A world free of drunkenness, violence, and poverty. A world where the lower classes could be bred out of existence, no longer a burden to society.

In 1869, Francis Galton published *Hereditary Genius,* outlining his plans for a better tomorrow. Galton argued that the British government should issue certificates of fitness to worthy young men and women and offer money for every child produced. And it wouldn't be expensive. Galton believed that if British citizens were willing to spend "only one-twentieth of the cost spent for the improvement of the human race that is spent on the improvement of the breed of horses and cattle, what a galaxy of geniuses might we not create!" He called his plan *eugenics,* from the Greek for "well born."

But there was a darker side. "I do not, of course, propose to neglect the sick, the feeble or the unfortunate," wrote Galton, "but

I would exact an equivalent for the charitable assistance they receive *by preventing the more faulty members of the flock from breeding*." Galton argued that lunatics, criminals, and paupers should be placed in monasteries and convents "for the purpose of restricting their opportunities for producing low-class offspring." Breeding had become weeding.

In the early 1900s, eugenics crossed the ocean and landed in a small cove near Huntington, New York. Here, on the beaches of Long Island, America's eugenics movement took hold. The man who championed Galton's cause was Charles Davenport, an accredited member of America's academic elite. The son of a long line of English and colonial New England Congregationalist ministers, Davenport had received his doctorate in zoology from Harvard before teaching at the University of Chicago. In 1904, he was appointed director of the Station for Experimental Study of Evolution at Cold Spring Harbor.

Charles Davenport worshipped Francis Galton. "As [society] claims the right to deprive the murderer of his life," said Davenport, "so also it may annihilate the hideous serpent of hopelessly vicious protoplasm." Davenport argued that the cost of taking care of America's defective citizens was about $100 million a year. It was time to do something about it. So he created the Eugenics Record Office at Cold Spring Harbor, keeping careful tabs on those who were worthy and those who weren't. A few years later, he plucked Harry Laughlin out of a desolate, one-room schoolhouse in Livonia, Missouri, and appointed him superintendent. Davenport would be the researcher (supplying scientific "evidence" to support their cause), and Laughlin would be the lobbyist persuading those in power to pass laws to eliminate this lesser breed of citizens).

In October 1910, the Eugenics Record Office opened for business. Its mission was clear: Determine which Americans were of inferior stock and prevent them from marrying or having children. The first step was to confine them to unisex institutions for those deemed insane or feebleminded. And later to sterilize those who were still roaming free.

Determining who would be targeted wasn't going to be easy. Davenport asked his team of field workers to create family trees for unwanted traits that, according to him, lay "hidden in records of our 42 institutions for the feebleminded, our 115 schools and homes for the deaf and the blind, our 350 hospitals for the insane, our 1,200 refuge homes, our 1,300 prisons, our 1,500 hospitals, and our 2,500 almshouses." Davenport's plan included not only those who were unfit, but also those who had someone in their family who might have been unfit. He needed to eliminate their bloodline from America's gene pool. No stone could be left unturned. To store his valuable data, he built a fireproof vault.

As a first order of business, Davenport and Laughlin published their top ten list of "degenerate protoplasm": (1) the feebleminded; (2) the poor; (3) alcoholics; (4) criminals; (5) epileptics; (6) the insane; (7) the "constitutionally weak"; (8) those suffering from venereal diseases; (9) the deformed; and (10) the deaf, blind, or mute. (No attempt was made to distinguish those with blurry vision from those who were blind or those with bad hearing from those who were deaf.) Davenport and Laughlin calculated that their program would include about a million Americans currently in the state's care, three million who weren't in the state's care, and seven million family members. These 11 million people—according to the Eugenics Record Office—represented the bottom tenth of the U.S. population. The time had come to prevent them from reproducing.

Davenport and Laughlin could easily determine who was a criminal from jail records; who was blind or deaf from eye and ear examinations; and who had venereal diseases from hospital and clinic records. But how were they going to accurately determine who was feebleminded? Fortunately, a European researcher had made the task a lot easier.

In the early 1900s, a French psychologist named Alfred Binet created an intelligence test. A few years later, the test was modified by a Stanford researcher and renamed the Stanford-Binet test. Now the eugenicists had a hard and fast number they could rely on: 70. They determined that anyone with an intelligence quotient (or IQ) score of less than 70 was unfit for procreation. To celebrate the moment, they created a new word: "moron," from the Greek *moros* meaning "stupid" or "foolish." Not everyone was celebrating. Walter Lippmann, a syndicated columnist, wrote in the *New Republic* that the IQ test represented "a new chance for quackery in a field where quacks breed like rabbits." Most Americans, however, were excited about the chance to eliminate the least among them. And the IQ test was clear and objective, a good place to start. Lippmann's comments were ignored.

Eugenicists had completely bastardized Mendel's laws. Although it was true that physical characteristics like eye color could be mapped to single genes, such was not the case with traits like criminality, alcoholism, epilepsy, deafness, or susceptibility to venereal diseases. Not everything could be accounted for by strict Mendelian genetics. Nonetheless, the false notion that selective breeding could make for a better society would soon allow Americans to cloak some of their worst prejudices in the gilded robes of science.

In retrospect, given the absurdity of eugenics and its goals, one could only imagine that its pursuit would have been relegated to underfunded cranks working without public or mainstream scientific support. In fact, the opposite was true.

From its inception, the Eugenics Research Office had an advisory board that read like a who's who of academic royalty. The list included Alexis Carrel, a Nobel Prize–winning surgeon from the Rockefeller Institute; William Welch, a world-renowned pathologist from Johns Hopkins School of Medicine and future president of the American Medical Association; Stewart Paton, a psychiatrist from Princeton University; Irving Fisher, a public affairs professor from Yale University; James Field, a political economist from the University of Chicago; and three professors from Harvard—physiologist W. B. Cannon, immigration expert Robert DeCourcy Ward, and neuropathologist E. E. Southard.

And, far from being underfunded, the Eugenics Record Office was awash in funds, having been awarded tens of millions of dollars from the Carnegie Foundation (steel), the Rockefeller Institute (oil), Mrs. E. H. Harriman (rail), and George Eastman (photography). The State Department, the Army, and the Departments of Agriculture and Labor also supported Davenport and Laughlin.

In addition, eugenics was embraced by some of the world's most prominent, most influential, and most respected citizens.

David Starr Jordan, the president of Indiana University and founding president of Stanford University, was the first academic to popularize eugenics in his book, *Blood of the Nation.*

Alexander Graham Bell, who had invented the telephone and performed pioneering research in hearing loss, prepared a form used by eugenicists to document deafness.

H. G. Wells, the British novelist best known for his books *The Time Machine* and *The War of the Worlds,* wrote, "We want fewer and better children . . . and we cannot make the social life and the world peace we are determined to make with the ill-bred, ill-trained swarms of inferior citizens that [have been] inflict[ed] upon us."

Margaret Sanger, the founder of the American Birth Control League, tirelessly promoted a union between a woman's right to choose and eugenics. As a nurse, Sanger had been sickened by the inability of the poor to prevent unwanted births. She argued that birth control would allow for "more children from the fit and less from the unfit." It was time, Sanger said, for "human weeds to be extirpated."

John Harvey Kellogg, who operated a health sanatorium that offered fanciful foods for the wealthy, founded the Race Betterment Foundation of Battle Creek, Michigan. Eight years after he invented the cornflake, Kellogg said, "We have wonderful new races of horses, cows, and pigs. Why should we not have a new and improved race of men? A race of human thoroughbreds." Espousing a common belief of his time, Kellogg said that those destined for abnormality had been "begotten in lust."

George Bernard Shaw, an Irish playwright and a founder of the London School of Economics, also embraced eugenics. The author of more than 60 plays, Shaw was probably best known for *Pygmalion,* which was later made into the musical *My Fair Lady*. Shaw is the only person to have won both a Nobel Prize in literature and an Academy Award. Despite his socialist leanings, Shaw wholly supported eliminating the underclass: "There is now no reasonable excuse for refusing to face the fact that nothing but a eugenic religion can save our civilization from the fate that has overtaken all previous civilizations." So much for Eliza Doolittle.

Theodore Roosevelt also weighed in. On January 3, 1913, Roosevelt sent a letter to Charles Davenport. "Some day," he wrote, "we will realize that the prime duty of the good citizen of the right type is to leave his or her blood behind him in the world; and that we have no business to permit the perpetuation of citizens of the wrong type."

Although Pope Pius XI would later speak out against eugenics, most American clergy backed the efforts of the Eugenics Record Office, citing Matthew 7:16: "Are grapes gathered from thorn bushes, or figs from thistles?" Dr. Albert Wiggam, an author and leading member of the American Association for the Advancement of Science, also believed that eugenics was divine. "Had Jesus been among us," said Wiggam, "He would have been president of the First Eugenic Congress."

THE ZEALOUS EFFORTS of Davenport and Laughlin shaped a nation.

By 1928, about 400 colleges and universities in the United States offered courses in eugenics, and 70 percent of all high school biology textbooks embraced the pseudoscience. Eugenicists sponsored "Fitter Families" competitions and traveled to state fairs, Kiwanis conventions, PTA meetings, museums, and movie theaters. One exhibit, titled "Some People Are Born to Be a Burden on the Rest," featured a series of blinking lights. One light, which flashed every 48 seconds, indicated the birth of a "defective person"; another, which flashed every 50 seconds, indicated that someone had just been sent to jail and that "very few normal people ever go to jail"; a third, which flashed only every 7 minutes, indicated the birth of a "high-grade person." The exhibit explained that "every 15 seconds $100 of your money goes for the care of persons with bad heredity."

With the support of wealthy philanthropists, influential citizens, and respected academics, the eugenics movement in the United States changed the law. Four states prohibited the marriage of alcoholics, 17 prohibited the marriage of epileptics, and 41 prohibited the marriage of those deemed feebleminded or insane. By the mid-1930s, America was the world leader in banned marriages. (Marriage restriction laws weren't declared unconstitutional until 1967.)

WITH AMERICANS TAKING THE LEAD, eugenics became an international phenomenon.

In 1912, the First International Congress of Eugenics took place in London. Alexander Graham Bell was the honorary president. Scientists from the United States, Belgium, England, France, Italy, Japan, Spain, Norway, and Germany attended. Nine years later, the Second International Congress of Eugenics was held in New York City. A prominent American eugenicist named Henry Fairfield Osborn gave the keynote address. "As science has enlightened government in the prevention and spread of disease," he said, "it must also enlighten government in the prevention of the spread and multiplication of worthless members of society." Of the 53 papers presented at the meeting, 42 were from American researchers. Despite its international appeal, eugenics was an American science.

In 1917, eugenics entered the popular culture with the release of the Hollywood movie *Black Stork*. Promoted as a "eugenics love story," the movie featured a "defective" child who was allowed to die. The message of *Black Stork,* and the advertisements promoting it, were clear: Kill the defectives, save the nation. The movie played to enthusiastic audiences for more than a decade.

With the popularity of *Black Stork* and the support of lawmakers, American citizens were ready to take the next step—to legislate forced sterilization. These procedures had the blessing not only of the medical and scientific communities, but also eventually of the United States Supreme Court. Eugenicists argued that the country would need to sterilize the lower 10 percent of the population and to continue to sterilize the lower 10 percent until the gene pool was pure. Their initial goal was to sterilize 14 million Americans. When the dust settled, 65,370 poor, syphilitic, feebleminded, insane, alcoholic, deformed, lawbreaking, or epileptic Americans in 32 states had been sterilized. California alone was responsible for more than 20,000 of them. Few, if any, Americans rose in protest. It was one of the darkest moments in American history.

Most of those sterilized didn't understand what was being done, surprised that they could no longer bear children later in life. Some were told they were having a different surgical procedure. (Because of its popularity in the South, sterilizations were often referred to as "Mississippi appendectomies.") Others were told to sign a form that they couldn't read. In 1927, civil libertarians were delighted when the United States Supreme Court agreed to hear the case of a woman who was being sterilized against her will. At last, the most disenfranchised members of society would have their day in court. The person who was being sterilized was Carrie Buck. The doctor who was to perform the sterilization was John Bell. The case, one of the most famous in the history of American jurisprudence, was called *Buck* v. *Bell.*

On July 3, 1906, Frank and Emma Buck gave birth to a daughter, Carrie. When Frank deserted the family, Emma turned to prostitution. On April 1, 1920, Emma was forced to admit to a eugenics commission that she was a prostitute who had contracted syphilis. As a consequence, she was sent to the Virginia State Colony for Epileptics and Feebleminded in Lynchburg, where she would remain for the rest of her life. Carrie, who was three years old at the time, was sent to a foster home. All through elementary school Carrie showed herself to be an enthusiastic and able student; nonetheless, her foster parents took her out of school in the sixth grade to help with chores around the house. Later, she was lent out to other homes to help with their chores.

When Carrie Buck was 16, Clarence Garland, the nephew of her foster parents, raped her. Within a few months it was clear that she was pregnant. "He promised me that we would get married," said Carrie, "but we didn't." On January 23, 1924, embarrassed by the scandal, Carrie's foster parents had her committed to the colony in Lynchburg. Two months later, Carrie gave birth to a little girl, Vivian. While at Lynchburg, Carrie was given the Stanford-Binet test to determine whether she was subject to Virginia's new sterilization law. (Given its principal use, the Stanford-Binet intelligence test should have been called the Stanford-Binet feeblemindedness test.) Although Carrie was 17, she was said to function at the level of a 9-year-old and was labeled a moron.

With Carrie's test results in hand, Dr. John Bell, the superintendent of the Virginia colony, determined that she should be sterilized. Although about 80 people had already been sterilized in Virginia, eugenicists wanted to strengthen the state's law by testing it in court. On November 18, 1924, the Circuit Court of Amherst County heard the case. One of the first to testify was Harry Laughlin, who had traveled down from the Eugenics Record Office in Cold Spring Harbor. Laughlin said that Carrie was "immoral, untruthful, and a low-grade moron," even though he had never met her. At the time of the hearing, Carrie routinely read the newspaper and did the crossword puzzles. Laughlin said that Carrie's ancestors belonged to "the shiftless, ignorant, and worthless class of anti-social whites in the South," arguing that the Bucks were living proof of "Mendelian inheritance." Another ardent eugenicist said that Carrie's sterilization would "raise the standard of intelligence in the state." The social worker who had examined six-month-old Vivian Buck also testified. "There is a look about it that is not quite normal," she said. "But just what it is, I can't tell." The district court was impressed, ordering Carrie's sterilization.

On November 12, 1925, the Virginia Supreme Court of Appeals supported the verdict.

In September 1926, the United States Supreme Court accepted the case of *Buck* v. *Bell* for review. The Chief Justice was former president William Howard Taft. But the man who wrote the opinion for the majority wasn't Taft; it was Oliver Wendell Holmes, Jr., one of the clearest-thinking, most respected jurists in the land. A proud defender of the Constitution and individual liberties, Holmes had authored nearly a thousand valued opinions. (One contained a phrase that is still used today. Regarding an individual's First Amendment right to speak freely and without restraint, Holmes wrote, "the most stringent protection of free speech would not protect a man from falsely shouting fire in a theater and causing panic.") At the time of *Buck* v. *Bell,* Oliver Wendell Holmes, Jr., a veteran of the Civil War, was 86 years old.

During the trial, Carrie Buck's lawyer made an ominous prediction about what would happen if forced sterilizations were allowed to proceed. "A reign of doctors will be inaugurated in the name of science," he warned, "*even races* may be brought within the scope of such regulation, and the worst forms of tyranny practiced." The Court was unmoved. On May 2, 1927, by a vote of 8 to 1, justices ruled in favor of Carrie Buck's sterilization. Even Louis Brandeis, the Court's most liberal justice, sided with the majority. Holmes, an enthusiastic eugenicist, wrote the opinion: "Carrie Buck is a feeble-minded white woman. She is the daughter of a feeble-minded mother in the same institution, and the mother of an illegitimate feeble-minded child. It is better for all the world, if instead of waiting to execute degenerate offspring for crimes, or to let them starve for their imbecility, society can prevent those who are manifestly unfit from continuing their kind." Then Oliver Wendell Holmes, Jr., wrote the words that placed

*Buck* v. *Bell* in the pantheon of America's most embarrassing Supreme Court decisions: "Three generations of imbeciles are enough," he wrote, effectively solidifying laws that even the most ardent eugenicists thought were unenforceable. One critic later wrote that Holmes's opinion represented "the highest ratio of injustice per word ever signed on by eight Supreme Court Justices." (The United States Supreme Court has never officially overturned the verdict in *Buck* v. *Bell*.)

On October 19, 1927, her legal options exhausted, Carrie Buck was sterilized; she thought she was having an appendectomy. Twenty years later, the United States Supreme Court's verdict in *Buck* v. *Bell* would be presented in support of SS officer Otto Hofmann during the Nuremberg Military Tribunal investigating Nazi war crimes.

BEFORE 1916, AMERICAN EUGENICISTS had focused on individuals and their families. It was all about bloodlines and pedigrees. But in 1917—with the passage of the first of a series of restrictive immigration laws—the focus began to change. And when it did, the stage was set for a level of evil that was unprecedented and will likely remain forever unmatched.

The person responsible for this shift in thinking was a New York City lawyer and conservationist named Madison Grant. In 1916, Grant wrote *The Passing of the Great Race*. Framed as a scientific treatise, Grant made the case that Americans were committing what he called "race suicide." Undesirable traits weren't just shared among certain families; according to Grant's book, they were shared among certain races. If Americans really wanted to purify the gene pool, they needed to prohibit the entry of undesirable races into their country. Grant argued that America needed to become America again. And that the only way this was going to happen was if we removed the weeds and allowed people of Grant's race to flourish.

A decade later, when Madison Grant's book was translated into German, no one would embrace his notion of race purity more than a young soldier imprisoned in a fortress in Landsberg.

MADISON GRANT WAS BORN ON November 19, 1865, in the exclusive Murray Hill section of New York City. His mother was descended from the first band of colonists to settle in the New Netherland who, after securing land grants on Manhattan Island, founded the city of New Amsterdam (later called New York City). His father was descended from the first Puritan settlers of New England, whose family included a colonial governor of Connecticut and the founder of Newark, New Jersey. During the Civil War, Grant's father won the congressional Medal of Honor, the highest military award given for bravery in the United States.

Throughout his youth, Grant was educated by private tutors. When he was 16—to complete his classical education—he was sent to Dresden, Germany. Once back in the United States, he applied to Yale University, where he was grilled for three days in mathematics, German, Greek, and Latin. He passed with flying colors. Later, Grant attended Columbia Law School, opened his own law practice, and joined the elite clubs in New York City, interacting with some of the nation's most powerful men. Affable, charming, considerate, soft-spoken, well liked, and with little interest in practicing law, Grant turned his attention to his first love: conservation.

Before he wrote *The Passing of the Great Race,* Madison Grant was the single most effective conservationist in America. He founded the Bronx Zoo as well as the Wildlife Conservation Society, which designed zoos in Queens, Prospect Park, and Central Park as well as the New York Aquarium. Grant singlehandedly saved the American bison from extinction and played a key role in creating Denali

National Park in Alaska, Everglades National Park in Florida, Olympic National Park in Washington, and Glacier National Park in Montana. He also devoted himself to saving whales, bald eagles, and pronghorn antelopes. When Grant launched his campaign to preserve America's wilderness, Yellowstone National Park in Wyoming was called "The National Park." At the time of Grant's death, and due largely to his efforts, a vast system of national parks stretching over more than eight million acres provided refuge to tens of thousands of large game animals.

Perhaps Grant's greatest accomplishment followed a trip to northern California, where he saw the tallest living things on Earth: the California redwoods. When he first visited the redwoods in 1917, Grant witnessed trees that were more than 2,000 years old, alive at the time of Jesus. But Grant was sickened by the fact that many of these trees were being cut down for wood, so he founded the Save the Redwoods League, one of the most successful conservation efforts in the history of the United States, and a model for similar efforts that followed. Redwood National Park, created in 1968, was the culmination of his work.

When Madison Grant was coming of age, he could walk down the streets of New York City surrounded by people who, like him, had ties to colonial America; people who he believed were upright, who understood the rules of the republic and wanted to abide by them; people who had character. By the late 1880s, according to Grant, all of that had changed. The rate of immigration had doubled to more than half a million people every year. Worst of all, no longer were immigrants coming from countries in northwestern Europe like the British Isles, Scandinavia, and Germany, but they were coming from southern and Eastern Europe. Jonathan Spiro, in his book *Defending the Master Race,* described how Grant must have felt while strolling through his beloved city: "Grant felt increasingly beleaguered by the waves of swarthy immigrants engulfing his city. They were filling up almshouses, cluttering the streets, and turning Manhattan into a dirty, lawless, turbulent cacophony of foreign barbarians . . . Grant was disgusted by what he saw as he braved the congested sidewalks of his native city. He was repulsed by the bizarre customs, unintelligible languages, and peculiar religious habits of the foreigners. As he was jostled by Greek ragpickers, Armenian bootblacks, and Jewish carp vendors, it was distressingly obvious to him that the new arrivals did not know this nation's history or understand its republican form of government." Grant's world was collapsing. He had to do something to conserve America for natives like himself—to make America America again.

Of all the new immigrants, no group drew Grant's attention more than the Jews. Between 1880 and 1914, one-third of Eastern Europe's Jews immigrated to the United States. The Jewish population in New York City, which numbered about 80,000 in 1880, was more than a million only 30 years later; half were packed into an area of only 1.5 square miles in New York's Lower East Side—a population density greater than any other city in the world, including Bombay.

Madison Grant had a label for people who, like himself, were descended from Scandinavia and Germany; he called them Nordics. And, like his efforts to preserve the American bison or the California redwood, he wanted to preserve America's Nordic race. This was the theme of what was to become his best-selling book. (What American eugenicists called the Nordic race, Europeans called the Aryan race.) In the end, Grant's campaign to save his America would be an excuse for the homophobia, misogyny, anti-Semitism, and anti-Catholic sentiments that were so common in the 1920s—most ardently expressed by the Ku Klux Klan, a major political force in the South with more than five million members. Madison Grant would provide a scientific basis for their prejudices, as well as for the prejudices of a rising National Socialist Party in Germany.

In the spring of 1916, Charles Scribner's Sons published *The Passing of the Great Race*. The book was reprinted in 1922, 1923, 1924, 1926, 1930, 1932, and 1936, selling more than 1.6 million copies—one of the most popular scientific treatises in history. In it, Grant explained that genes determined character and that character determined history. He proposed three scientific "facts":

1. The human species is divided into biologically distinct races, with the Nordic race at the top.
2. The intellectual, moral, and temperamental traits of each race are unaffected by the environment. (Nature is everything. Nurture is irrelevant.)
3. If a member of an inferior race mates with a member of a superior race, the result is a member of the inferior race. "The cross between a white man and an Indian is an Indian," wrote Grant. "The cross between a white man and a Negro is a Negro; the cross between a white man and a Hindu is a Hindu; and the cross between any of the three European races and a Jew is a Jew." (This last phrase provided the basis for a law that would later be passed in Nazi Germany.)

Grant had perverted Gregor Mendel's experiments on peas to an explanation of European history. Mendel had taken pea plants with green leaves and painted them yellow. He wanted to see if any of the painted plants had offspring with yellow leaves. They didn't. Genes were everything. Grant used this finding to advance his notion that genes were inviolate, immutable—that what's bred in the bone will always come out in the flesh. "It has taken us fifty years to learn that speaking English, wearing good clothes and going to school and to church does not transform a Negro into a white man," he wrote in *The Passing of the Great Race.* "Nor was a Syrian or an Egyptian freedman transformed into a Roman by wearing a toga and applauding his favorite gladiator in the amphitheater. Americans will have a similar experience with the Polish Jew, whose dwarf stature, peculiar mentality and ruthless concentration on self-interest are being engrafted upon the stock of the nation."

To Madison Grant, it was easy to tell who was Nordic and who wasn't. All you had to do was look. Nordic people had "wavy brown or blond hair and blue, gray or light brown eyes, fair skin, [and a] high narrow and straight nose, [all of] which are associated with great stature and a long skull as well as with abundant head and body hair." These characteristics, according to Grant, could easily be found in some of the world's greatest paintings. "It would be difficult to imagine a Greek artist painting a brunette Venus," he wrote. "In church pictures, all angels are blond, while the denizens of the lower regions revel in deep brunetness. In depicting the crucifixion, no artist hesitates to make the two thieves brunet in contrast to the blond Saviour." Jesus of Nazareth, apparently, was Nordic.

Nordics were hunters, sailors, explorers, painters, soldiers, and kings—the very best the human species had to offer. According to Grant, Alexander the Great was Nordic; so were Dante, Raphael, Titian, Michelangelo, da Vinci, Sophocles, Aristotle, and even King David (because of biblical references to his "fairness").

Americans loved Madison Grant's book, which was praised by the *Yale Review,* the *American Historical Review,* the *New York Herald,* the

*Nation,* the *New York Sun,* and *Science.* Presidents Herbert Hoover and Theodore Roosevelt were both impressed by Grant's rigor and insight; Roosevelt even sent him a letter. "This book is a capital book," he wrote. "In purpose, in vision, in grasp of facts our people most need to realize." Calvin Coolidge, also taken by the book, said that America must cease to become "a dumping ground for advancing hordes of aliens."

Grant's theories appeared in poems, paintings, scientific journals, and ladies' magazines. Margaret Sanger included quotes from Grant's book in her speeches. In 1924, when Clarence Darrow stood up to defend Nathan Leopold and Richard Loeb—two Jewish students at the University of Chicago who had kidnapped and killed a 14-year-old boy—he argued that bad genes had been responsible for their crimes. At the same time, Hiram Wesley Evans, Imperial Wizard of the Ku Klux Klan, included quotes from Grant's book in his white supremacist pamphlets.

Grant's book also contained a fourth scientific "fact": Only eugenics could preserve the Nordic race. "This is a practical, merciful, and inevitable solution to the whole problem," he wrote, "and can be applied to an ever widening circle of social discards . . . and perhaps *ultimately to worthless race types.*" Grant had used the words "practical," "merciful," and "inevitable" in describing his solution to the problem. In Nazi Germany, the phrase that would later emerge was "final solution."

Grant's book ended with a plea for his kind of America: "We Americans must realize that the altruistic ideals, which have controlled our social development during the past century, and the maudlin sentimentalism that has made America 'an asylum for the oppressed,' are sweeping the nation toward a racial abyss. If the Melting Pot is allowed to boil without control and we continue to follow our national motto and deliberately blind ourselves to all 'distinctions of race, creed or color,' the type of native American of Colonial descent will become as extinct as the Athenian in the age of Pericles and the Viking in the days of Rollo." Grant's lament was in direct contradiction to the poem emblazoned on the Statue of Liberty written by Emma Lazarus, an American Jew.

> Give me your tired, your poor,
> Your huddled masses yearning to breathe free,
> The wretched refuse of your teeming shore.
> Send these, the homeless, tempest-tost to me.

ALTHOUGH MADISON GRANT'S BOOK swept a nation, not everyone bought into his deception.

Thomas Hunt Morgan, a geneticist who would later win the Nobel Prize for his work on chromosomes, noted that there was no such thing as the Nordic race or the Aryan race. Biologically speaking, all humans were products of an intermixture of many genetic backgrounds. There was only one race: the human race.

Emily Greene Balch, a Wellesley economist who would also later win the Nobel Prize, saw eugenics as just another sad example of the powerful exploiting the weak: "Rash is the man who passes lightly from skull measurements to vast unprovable sociological and historical generalizations. The pseudoscience [that] makes it the function of the strong man to purge the world of the weak, might, one hoped, by this time have gone out of date."

H. L. Mencken, a satirist, essayist, and editor of the *American Mercury,* was sickened by the hauteur and superiority of those who had been born on third base and thought they had hit a triple: "My impression, though I am blond and Nordic myself, is that the genuine member of that great race, at least in modern times, is often indistinguishable from a cockroach."

Perhaps the most withering dissent came from the pen of G. K. Chesterton, a British author and poet. Regarding pending immigration laws, Chesterton drove a wedge between the science of Gregor Mendel and the pseudoscience of Madison Grant: "One does not need to deny heredity in order to resist such legislation any more than one needs to deny the spiritual world in order to resist an epidemic of witch burning." Unfortunately, the voices of Morgan, Balch, Mencken, and Chesterton were lost in the cacophony of support for Madison Grant and his theories.

In the end, Madison Grant's theories were refuted by history. Grant had predicted that it would "take centuries" for immigrants to assimilate into the American culture. It took one generation. European immigrants quickly lost their accents, earned their degrees, and rose to prominent positions in business, medicine, and the law. Environment, as it turned out, mattered.

AFTER PUBLISHING *The Passing of the Great Race,* Madison Grant was America's premier race theorist in a country that was perfectly willing to believe that race was everything. During the next decade, Grant used his influence over Congress to encourage the passage of four immigration laws designed to make America more American. One scholar at the time referred to them as "America's most ambitious program of biological engineering."

In 1917—one year after Madison Grant published his book—Congress passed a law banning "all idiots, imbeciles, feebleminded persons, epileptics, insane persons [and] persons of constitutional psychopathic inferiority" from entering the country. The bill also mandated a literacy test. During deliberations, one congressman read directly from *The Passing of the Great Race*. As a consequence, about 1,500 immigrants a year were denied entrance. The tide was turning.

And no one was happier than Charles Davenport, who in a letter to Grant urged him to push forward on immigration restriction: "Can we build a wall high enough around this country so as to keep out these cheaper races; or will it be a feeble dam, leaving it to our descendants to abandon the country to blacks, browns, and yellows." A hundred years later, Donald Trump said, "People are pouring across our borders, which is horrible. We have to build a wall. I build some of the greatest buildings in the world. Building a wall for me is easy. And it would be a wall. It would be a real wall. Not a wall that people walk over."

In 1921, Congress passed the Emergency Quota Act, which further limited the number of immigrants. One congressman arguing in favor of the bill said, "The issue . . . is simply this: shall we preserve this country, handed down to us by a noble and illustrious ancestry, for Americans, and transmit it to our posterity as our forefathers intended; or shall we permit it to be overrun and submerged by a heterogeneous, hodgepodge, polyglot, aggregation of aliens, most of whom are the scum, the offal, and the excrescence of the earth." The year before the Emergency Quota Act passed, about 800,000 immigrants entered the United States; the year after it passed, that number was reduced to 300,000.

In 1924, Congress passed the Immigration Restriction Act, which imposed more restrictive quotas. Before World War I, as many as a million immigrants entered the United States every year. After 1924, that number was reduced to 20,000—a trickle that even the most ardent eugenicists could live with.

In 1929, Congress passed the National Origins Act, further restricting immigration. Eugenicists had accomplished exactly what they had wanted. More immigrants came into the United States in 1907 alone then entered during the next quarter century. Madison

Grant was thrilled. "[This is] one of the greatest steps forward in the history of this country," he said. "We have closed the doors just in time to prevent our Nordic population from being overrun by the lower races." The director of Ellis Island, the entry point for most European immigrants, commented that immigrants were now starting to look more like Americans.

Perhaps Madison Grant's most cynical alliance was with Marcus Garvey, an African American. Garvey wanted blacks to take pride in their race, pride in their accomplishments; he didn't want them to feel compelled to assimilate into society. Garvey condemned interracial marriage, preached racial purity, and yearned for an African homeland. By 1920, when blacks were routinely being lynched in the South, Garvey's Back-to-Africa campaign had two million members. Where Marcus Garvey sought a homeland for men and women who were treated poorly because of the color of their skin, Grant sought deportation of what he believed to be a subspecies of humans who were poisoning the gene pool. The eugenics movement produced no pairing sadder than this one.

In 1925, Madison Grant's *The Passing of the Great Race* was translated into German where it was read by a disgruntled corporal who had recently been sent to prison for his part in a riot against the government in Bavaria: Adolf Hitler. After reading the book, the 36-year-old revolutionary sent a fan letter to Grant: "This book is my Bible," he wrote. During his nine months in prison, Hitler had read several books by American eugenicists, calling his prison stay "his university." Hitler would soon launch a national movement that would forever damn the field of eugenics to the lower reaches of hell. But, despite popular belief, what was about to happen in Germany didn't start on a rallying stand in Munich; it started in a law office in New York City.

As Hitler sat in Landsberg Prison, he worked on his autobiographical manifesto *Mein Kampf (My Struggle).* The first volume was published in 1925, the next, in 1926. To say that Madison Grant's *The Passing of the Great Race* had influenced Adolf Hitler's *Mein Kampf* would be an understatement; in some sections, Hitler had virtually plagiarized Grant's book. For example, in *The Passing of the Great Race,* Grant wrote, "It has taken us fifty years to learn that speaking English, wearing good clothes and going to school and to church does not transform a Negro into a white man." In *Mein Kampf,* Hitler wrote, "But it is a scarcely conceivable fallacy of thought to believe that a Negro or a Chinese, let us say, will turn into a German because he learns German and is willing to speak the German language." In 1936, three years after Adolf Hitler came to power, the Nazi Party listed Madison Grant's *The Passing of the Great Race* as essential reading.

Francis Galton, Charles Davenport, Harry Laughlin, Madison Grant, and Adolf Hitler all shared several features: All were, by their definition, Nordic; all believed that Nordics should procreate freely while non-Nordics should be prevented from procreating; and all were childless.

In 1933, the year that he came to power, Adolf Hitler passed the Law for the Prevention of Hereditarily Diseased Offspring. The list of those to be sterilized was virtually identical to that first generated by the Eugenics Record Office in Cold Spring Harbor. Clinics were established and doctors were fined if they didn't comply with the law. Within a year, 56,000 Germans had been sterilized; by 1935, 73,000; by 1939, 400,000, dwarfing the number of sterilizations performed in the United States. The procedure was so common that it had a nickname: *Hitlerschnitte,* "Hitler's cut." Americans took note. Joseph DeJarnette, superintendent of Virginia's Western State Hospital, lamented, "Hitler is beating us at our own game!"

Then Hitler moved from sterilization to murder. Handicapped children in hospitals were starved, injected with lethal drugs, or—in a tribute to ancient Sparta—exposed to the cold. Initially, only grossly deformed newborns were killed. Then killing of the unfit extended up to 3 years of age, then 8, then 12, then 16. Then the definition of "handicapped" broadened to include anyone with an incurable disease or with learning difficulties. Even chronic bedwetters were at risk. Under the auspices of Karl Brandt, Hitler's personal physician, Germany's euthanasia program soon extended to the elderly, infirm, insane, and incurably ill. More than 70,000 German adults were killed, initially by lethal injection, and eventually by mobile gas chambers that traveled from clinic to clinic. German physicians sanctioned each and every one of the killings. (When Karl Brandt was tried for war crimes in Nuremberg, and later sentenced to death, he offered *The Passing of the Great Race* as an exhibit in his defense.) Adolf Hitler's Germany had become the embodiment of "the reign of doctors" that Carrie Buck's lawyer had predicted during his pleadings before the United States Supreme Court.

In 1935, Adolf Hitler passed the Nuremberg Laws, stripping Jews of their rights as citizens as well as outlawing sexual relationships or marriage between Jews and Aryans. The Eugenics Record Office praised the Nuremberg Laws as sound science. Eventually, Jews were segregated into ghettos and sent to concentration camps for what Hitler termed *Die Endlösung*—the "final solution." "We shall regain our health only by eliminating the Jew," he said. At least six million Jews, Slavs, Romany, homosexuals, and "mental defectives" were murdered. Madison Grant's "race suicide"—a fear that his Nordic race would be diluted out by inferior races—had become ethnic genocide. "National Socialism is nothing but applied biology," said Rudolf Hess, Hitler's deputy führer.

AMERICAN EUGENICISTS EMBRACED Hitler's efforts. Both the Carnegie Institute and Rockefeller Foundation supported a German scientific establishment committed to sterilization and euthanasia. Indeed, IBM provided machinery to help the Nazis sort out family pedigrees to determine who was Jewish and who wasn't.

*Eugenical News,* the official voice of the American eugenics movement, wrote, "[M]ay we be the first to thank this *one* man, Adolf Hitler, and to follow him on the way towards biological salvation and humanity."

On February 12, 1935, C. M. Goethe, a member of the board of trustees of a eugenics group called the American Betterment Foundation, wrote a letter to a foundation worker: "You will be interested to know that your work has played a powerful part in shaping the opinions of the group of intellectuals who are behind Hitler in his epoch-making program. I want you, my dear friend, to carry this thought with you for the rest of your life, that you have really jolted into action a great government of 60 million people."

Editorials in mainstream scientific publications like the *Journal of the American Medical Association,* the *American Journal of Public Health,* and the *New England Journal of Medicine* also supported the efforts of Adolf Hitler, the world's most effective eugenicist.

To be fair, American eugenicists had yet to observe or to frankly imagine the horrors exacted behind the walls of German concentration camps. When they did, eugenics would become an obscenity. But before that could happen, one more scene had to play out near a small industrial town in southern Poland: Auschwitz. Here, between May 1943 and January 1945, eugenics made its fanatical last stand.

IN THE 1940S, Germany's most influential eugenics scientist was Dr. Otmar Freiherr von Verschuer, head of anthropology at the Kaiser

Wilhelm Institute for Heredity Biology and Racial Hygiene in Dahlem. Verschuer studied Jews. Much to Hitler's delight, Verschuer found that Jews suffered disproportionally from a variety of diseases including diabetes, flat feet, deafness, and nervous conditions. In 1936, Verschuer's findings were reported and praised in *Eugenical News*.

To Verschuer and the Nazis, Jews weren't an ethnic group; they were a biological entity, one that could be distinguished from the general population by their physical characteristics. One Verschuer protégé charged with determining characteristics that were uniquely Jewish was an eager young doctor who studied dimples, jawlines, and ear dimplings. His name was Josef Mengele.

MENGELE WAS BORN ON March 16, 1911, to a wealthy family in Günzburg, Germany, that made farm machinery. (Today, the company is the third largest manufacturer of threshers in Germany; all equipment still proudly bears the name MENGELE.) After receiving his medical degree from the University of Frankfurt in 1938, Mengele traveled to the front, eventually returning to Germany to further his eugenic studies. On May 30, 1943, he arrived at the Auschwitz-Birkenau concentration camp where more than 100,000 prisoners awaited him.

When Jews were first lined up on the unloading docks at Auschwitz, they invariably heard the following command from German officers walking up and down the lines: *Zwillinge! Zwillinge!* ("Twins! Twins!"). Because twins were genetically identical, they were perfect for genetic studies. Mengele wanted to find ways to build a master race: one free of disease and capable of transmitting the best Aryan traits. In the two years he was at Auschwitz, he studied 1,500 pairs of twins. His fellow officers called them "Mengele's Children."

Mengele's studies began by taking the children to Barrack 14, Camp F, the "Twin Camp." There he would strip them naked, take photographs, and carefully measure and record every possible physical characteristic. Then he put a syringe into their veins to test their blood, and needles into their backs to test their spinal fluid. Later, he performed a series of experiments that brought eugenics to its final, hideous end. When he found one twin who sang well and another who didn't, Mengele operated on their vocal cords; one of the brothers never spoke again. He forced twin girls to have sex with twin boys to see if they would produce twins. To create Aryan features artificially, he injected a Nordic blue dye into the eyes of children, leaving many blind. He took one hunchbacked child and connected the veins in his wrists to the veins of his twin; then he connected them back-to-back. He wanted to see if he could transmit the misshapen spine from one child to another; following the surgery, the children couldn't stop screaming in horror. Their mother, who was able to procure a lethal dose of morphine, killed them both. Mengele thought that two Romany twins were infected with tuberculosis; when other German physicians in the camp disagreed, Mengele brought the children into a back room, shot them in the neck with his pistol, and performed an autopsy. "Yes, I dissected them while they were still warm," he told his colleagues, who had been right about their diagnosis. He infected children with typhus and tuberculosis to determine their susceptibilities to disease and performed mismatched blood transfusions to see what would happen. Mengele gave children electric shocks to see how much pain they could endure. He burned 300 children alive in an open fire. When children had heterochromatic eyes, he killed them and sent their eyes to Verschuer in packages marked, WAR MATERIALS: URGENT. Mengele asked one mother to tape up her breasts to see how long her newborn could survive without food. He dissected a one-year-old while the child was still alive. When the nightmare finally ended, fewer than 200 of the 3,000 children put into Mengele's care survived. And not a single piece of recognizable information was obtained. Josef Mengele and Adolf Hitler showed exactly what could happen when eugenics was put into the hands of narcissistic sadists with absolute power.

After the war, Mengele, who would later be called the Angel of Death, fled to Argentina, then Paraguay, then Brazil, where he drowned in São Paolo at the age of 68. Mengele saved the records from his experiments, certain that someday he would be hailed as a groundbreaking scientist. American eugenicists didn't share Mengele's sense of pride. After the war, the Eugenics Record Office at Cold Spring Harbor destroyed all of its records.

In 1952, a group of anthropologists, sociologists, geneticists, and psychologists gathered at the United Nations Educational, Scientific, and Cultural Organization (UNESCO) to put an end to Madison Grant's notion that race determined character and to the madness that it had wrought. They issued the following statements:

1. All men belong to the same species: *Homo sapiens.*
2. Race is not a biological reality but a social myth; the term should be dropped in favor of ethnic group.
3. There is no proof that the groups of mankind differ in their innate mental characteristics or intellectual capacity or that there is any connection between the physical and mental characteristics of human beings.

Although *The Passing of the Great Race* is still propagated by neo-Nazis and white supremacists on their websites, the book has now passed into history, unknown to most young students.

Madison Grant died on May 30, 1937, at the age of 72. As famous as he was in his time, his name has virtually disappeared. But it hasn't completely disappeared. Grant's name is still prominently displayed on a plaque at the base of the world's tallest living tree: the Founders' Tree in Redwood National Park. In 1991, the park's director, Donald Murphy, received an angry letter from a visitor, demanding that the plaque be removed and that the park stop honoring this loathsome man. Murphy wrote back: "[Madison] Grant was a creature of the nineteenth century and, as with many of his life contemporaries, he held beliefs that most of us, hopefully, find both absurd and abhorrent today. The sad truth is that [Grant] probably did not think too differently than many others who have been 'honored' for some historical role unrelated to the issue of race. I'm not sure that society can or should conduct a wholesale revision of history because the people of the past did not have a late-twentieth century vision of fairness and equality. As director of the California Department of Parks and Recreation, I don't ordinarily wear my ethnicity on my sleeve, so to speak, but in responding to your concerns I feel compelled to note that as an African American I think I have a personal perspective on the pain and suffering, the hurt and disappointment of racism."

Donald Murphy ended his letter with a statement that could itself have been put onto a plaque and nailed to the base of that tree. "Harmony among peoples comes from the true principles and attitudes of the present," he wrote, "not from purging the past."

The lesson here is little harder to describe but no less poignant: Beware of scientific biases that fit the culture of the time—**beware the zeitgeist**.

Imagine that a study has just been published in a prestigious medical journal claiming that a certain constellation of genes predisposes to violent behavior, like rape and murder. And that people living in Mexico are more likely to carry these genes. In all likelihood, several Republican presidential candidates of 2016 would have enthusiastically embraced this study. Now they would have had clear scientific evidence supporting what they had been saying all along—we need to restrict Mexican immigration and build a great wall to keep the Mexicans out; if not, a group of genetically inferior people will invade our country.

Although this might sound far-fetched, it's exactly what happened in 1916 with the publication of Madison Grant's *The Passing of the Great Race*. As a consequence, immigration slowed to a trickle. People then and now seem perfectly willing to ignore the fact that we all come from a common ancestor and are far more alike than different. There is no Nordic or Aryan or Mexican or Muslim or Syrian race. There's only one race: the human race.

When Lillian Hellman refused to participate in Senator Joseph McCarthy's communist witch hunt in the 1950s, her letter to the House Un-American Activities Committee contained a now famous quote: "I cannot and will not cut my conscience to fit this year's fashions." Hellman's comment should serve as a warning for all those who try to shoehorn scientific evidence into their cultural or political biases—advice that, as we'll discuss in the last chapter, remains unheeded today.

# CHAPTER 5

# TURNING THE MIND INSIDE OUT

*"Broken! Busted! Everybody has something to repair."*

—Billy Mays, American salesman

Between 1978 and 1991, a Milwaukee man named Jeffrey Dahmer killed 17 men and boys. Dahmer has been the subject of 25 books, hundreds of television programs, and thousands of newspaper and magazine articles. The public's fascination with Jeffrey Dahmer, however, had little to do with whom he'd killed and everything to do with how he'd killed them.

Dahmer was a special kind of serial killer. First, with the promise of $50 if they posed nude, he lured his victims to his apartment. Then he gave them drinks laced with sleeping pills. When they were unconscious, he strangled them, bludgeoned them, or cut their throats with a paring knife. Sometimes, before killing his victims, he bore a hole into their heads and injected hydrochloric acid or boiling water into the front of their brains, hoping to create, in his words, "zombie sex slaves."

On February 15, 1992, after deliberating for five hours, a jury found Jeffrey Dahmer guilty on 15 counts of murder. He was sentenced to 15 consecutive life terms totaling 957 years. Two years later, Christopher Scarver, a fellow inmate, beat Dahmer to death with a metal pipe.

Although most people know the story of Jeffrey Dahmer, they don't know that one of the atrocities performed in his chamber of horrors had—50 years earlier—won a Nobel Prize for its inventor.

IN AUGUST 1935, at a conference of neurologists in London, two physiologists from Yale, John Fulton and Carlyle Jacobsen, described a study they had done on two chimps: Becky and Lucy. Fulton and Jacobsen had taught the chimps to use sticks to get food that was out of reach. Sometimes the chimps got the food; sometimes they didn't. Lucy was consistently more patient than Becky. Where Lucy would just keep trying, Becky would fly into a rage, pull her hair, defecate, and throw her feces at the scientists.

The real experiment came next. Fulton and Jacobsen wanted to understand the role of specific areas of the brain in performing tasks that required memory. So they removed Lucy and Becky's frontal lobes (located just behind the forehead). Following the operation, Lucy no longer remembered how to get the food. The scientists concluded that Lucy's frontal lobes were responsible for synthesizing and storing recent memories. They also noticed something else. Becky still had trouble getting the food, but now she didn't care. "It was as though [she] had joined a happiness cult," said Jacobsen. John Fulton and Carlyle Jacobsen, it appeared, had invented a surgical treatment for anxiety.

Sitting in the audience was a Portuguese neurologist named Egas Moniz. Moniz was impressed by what he had just heard. He knew that many of his patients suffered from intense and often overwhelming anxiety. John Fulton remembered what happened next: "Dr. Moniz arose and asked, 'If frontal lobe removal prevents the development of experimental neuroses in animals, why would it not be feasible to relieve anxiety states in man by surgical means!?'" At first, Fulton, an experienced and well-respected neurologist, thought that Moniz was kidding. "At the time we were a little startled by the suggestion," he recalled, "for I thought that Dr. Moniz envisaged a bilateral lobectomy."

Moniz, however, wasn't thinking about a lobectomy, where the two frontal lobes were completely removed. Rather, he imagined a procedure where the frontal lobes would be cut off from the rest of the brain—something he would later call a *leucotomy* from the Greek *leuko,* meaning "white," referring to the white nerve fibers of the brain, and *tome,* meaning "knife." When Moniz's procedure crossed the Atlantic Ocean and entered the United States, it was called a *lobotomy*.

MONIZ WAS DETERMINED to extend Fulton and Jacobsen's experiments on chimps to people. First, he had to find a surgeon willing to do it. He settled on Almeida Lima, a neurosurgeon at the University of Lisbon. Within a few days, Moniz and Lima—without performing a single experiment on an animal and spending only an afternoon practicing the procedure on a cadaver—had picked their first patient.

On November 12, 1935, Almeida Lima drilled a hole into each side of the skull of a 63-year-old woman from a local insane asylum who suffered from crippling bouts of anxiety and paranoia. After drilling the holes, Lima—in a procedure mimicked by Jeffrey Dahmer decades later—injected half a teaspoon of alcohol into her frontal lobes. Then Lima closed up the holes. The operation took about 30 minutes. A few hours later, the woman was able to respond to simple questions. Two days later, she returned to the asylum and, according to Moniz, was much calmer—her anxieties and paranoia gone. Ecstatic, Moniz pronounced her cured.

Convinced that lobotomies worked, Moniz and Lima repeated the procedure on six more patients. Unfortunately, their technique lacked precision. Neither man felt comfortable that the alcohol they had squirted into the brain stayed in the frontal lobes. So they ordered a special surgical instrument from Paris: a long, thin rod, at the end of which was a wire loop. This allowed the team to remove small cores from the frontal lobes—like coring an apple. Within three months, their new lobotomy knife would be used on 13 more patients, bringing the total number to 20.

Staking his claim on the procedure, Moniz published a 248-page monograph describing his 20 patients: Seven were cured, seven were significantly improved, and six were unchanged. Psychosurgery was born. It was, said Moniz, "a great step forward." No longer did patients have to suffer from restlessness or bouts of anxiety, or from delusions or paranoia, or from mania or depression.

By the late 1930s, lobotomies had been performed in Cuba, Brazil, Italy, Romania, and the United States. Portugal, however, banned them. The psychiatrist who had initially referred patients to Moniz and Lima refused to provide any more. Soon, other Portuguese psychiatrists refused to provide patients. All had become horrified by the results. It was only later that the reasons for Portugal's ban became clear. By then, however, it was far too late.

When Egas Moniz attended the conference in London—and listened to John Fulton and Carlyle Jacobsen describe their chimp studies—it wasn't the first time that neurologists had learned about the workings of the brain's frontal lobes.

Perhaps no story was more instructive, more dramatic, or more unbelievable than that of Phineas Gage, a 25-year-old New England railroad worker who, on September 13, 1848, while preparing a hole for blasting powder, suffered a 3.5-foot iron spike driven through his face. The spike entered his cheek and exited through the top of his head, destroying his left frontal lobe. Miraculously, Phineas Gage lived for 11 more years. But, at least according to his friends, Gage was "no longer Gage." Before the accident, he was energetic, shrewd, and focused; afterward, he was ill-mannered, stubborn, and rude—once a responsible worker, now someone who couldn't hold down a job.

Patients with frontal lobe cancers had also been instructive. Like Phineas Gage, they became childish and apathetic, often dozed off to sleep, lacked initiative or will, lost their ability to plan ahead or make sound judgments, and had problems with attention, memory, language, and inhibition. One cancer patient, a 39-year-old New York City stockbroker named Joe A., was particularly fascinating. Following surgery for a frontal lobe tumor, Joe's memory seemed unaffected. Indeed, one group of neurologists examined him for an hour and couldn't find anything wrong. But Joe was a changed man. He lacked motivation to go back to work, was easily frustrated, spoke harshly about friends and neighbors, and most remarkably, became a hopeless braggart. While watching his son play baseball, Joe claimed that he himself was a better hitter than anyone he knew and that he would soon become a professional baseball player.

DESPITE HIS BOAST that "the intervention is harmless," Moniz's early lobotomy patients didn't do nearly as well as he had claimed. Patients often suffered from vomiting, diarrhea, incontinence, nystagmus (where the eyes rhythmically vacillate uncontrollably), ptosis (drooping of the upper eyelids), kleptomania, abnormal hunger, and a disturbed orientation of time and space. The Portuguese psychiatrist who had initially provided patients to Moniz and Lima later called the procedure "pure cerebral mythology."

Members of Sweden's Nobel Prize committee, however, either ignored or were unaware of any of these problems. In 1949, the committee recognized Egas Moniz "for his invention of a surgical treatment of mental illness." The *New York Times* immediately hailed the Nobel Prize winner as a brave explorer of the human brain: "Hypochondriacs no longer thought they were going to die; would-be suicides found life acceptable; [and] sufferers from persecution complexes forgot imaginary conspirators. Surgeons now think no more of operating on the brain than they do of removing an appendix." Lobotomies had entered the mainstream. Ironically, one country that never embraced lobotomies was Germany, believing them to be a violation of the Nuremberg Code, created to prevent doctors from performing the kind of cruel and unethical experiments that came to light following the Holocaust.

Within four decades of the Nobel Prize Committee's announcement, 40,000 lobotomies were performed throughout the world, more than half in the United States alone. America's love of lobotomies was due to the persistence and zealotry of one man—a man who had opened a Pandora's box of mental health treatments.

WALTER JACKSON FREEMAN WAS BORN ON November 14, 1895. Like Egas Moniz, Freeman was from a wealthy, prominent family. Freeman's father was a doctor, and his mother was the daughter of William Williams Keen, America's most famous surgeon. Although Freeman came to view his father as a second-rate surgeon whose financial dealings would eventually leave his family in ruin, he adored his grandfather. In William Keen, the young Freeman had much to live up to.

William Keen was the first surgeon in the United States, and one of the first in the world, to operate on a brain tumor. To accomplish this feat, he sprayed the entire operating room with carbolic acid, cut a hole in the man's skull, reached in with his ungloved hand, pulled out the tumor, sewed up the ripped blood vessels, and closed the hole with catgut. The operation was performed without the advantages of x-rays, blood transfusions, local anesthesia, or reliable lighting. Following the surgery, the man lived another 30 years. Keen was also the first surgeon to perform a colostomy, remarkable given that antibiotics hadn't been invented yet. He was the first surgeon to perform an end-to-end suture of a damaged nerve in a young boy's hand, allowing the child to continue to play the piano. He was the first surgeon to place a tube in the center of the brain to relieve a life-threatening buildup of spinal fluid. And he was the first surgeon to use open cardiac massage to save a patient's life. In 1921, Keen was also part of a team of physicians that diagnosed Franklin Delano Roosevelt's polio.

In short, Walter Freeman was never going to best his grandfather. But that didn't stop him from trying.

FREEMAN WAS THE ELDEST OF seven children. Growing up in a luxurious, three-story walk-up near Philadelphia's fashionable Rittenhouse Square, his childhood was eventful. When he was 14 months old, his grandfather removed 30 enlarged lymph nodes from his neck, leaving him with a permanent tilt of his head and droop of his shoulder. As a boy, he was one of the first to receive the newly made diphtheria antitoxin, imported from Germany, which saved his life.

As a young man, Freeman attended Philadelphia's prestigious Episcopal Academy followed by Yale University. After graduating from the University of Pennsylvania School of Medicine, he did his internship and residency in neuropathology at Philadelphia General

Hospital. Similar to many wealthy physicians in the early 1900s, Freeman extended his studies in Paris and Rome before returning to St. Elizabeth's Hospital in Washington, D.C., to become its director of laboratories. (St. Elizabeth's was one of the largest general hospitals in the United States, housing 4,000 staff and 7,000 patients. Charles Guiteau, who had assassinated President James A. Garfield, was a patient at St. Elizabeth's.) He also joined the faculty of both Georgetown and George Washington University medical schools. In 1928, Walter Freeman became the first chairman of George Washington University's Department of Neurology and Neurological Surgery.

Like his famous grandfather, Freeman soon earned the respect of his peers. He was elected to head the certification board for neurologists and psychiatrists and, sporting a jutting goatee, sombrero, and cane, was a dynamic and beloved lecturer. Freeman had a flair for the dramatic. While working at St. Elizabeth's, he took care of a sailor whose girlfriend had slipped a gold ring onto the sailor's penis during foreplay. When the sailor became erect, the ring got stuck. Freeman cut through the ring, twisted it off with a pair of pliers, fixed it, and kept it on his watch chain as a conversation starter (or conversation ender).

But where William Keen advanced the state of medicine in countless ways, Walter Freeman contributed virtually nothing. Freeman believed he could find structural differences in the brains of people with serious mental disorders. After examining more than 1,400 brains, he incorrectly concluded that patients with manic depression had anatomical differences depending on whether they were manic or depressed. Later, Freeman tried to visualize the brain by injecting a dye directly into its center, a dangerous procedure that would soon be abandoned. But Walter Freeman was unbowed. His mother, who had seen this streak of hauteur in her son from an early age, referred to him as "the cat who walks by himself" (a reference to Rudyard Kipling's *Just So Stories*).

In his efforts to best his famous grandfather, Freeman drove himself into the ground. To make time for what he believed would be the definitive textbook on neuropathology, he woke up at 4:00 a.m., wrote for three hours, drove to St. Elizabeth's where he worked until 5:00 p.m., then drove to his private practice, where he worked until 8:00 p.m. When he got home and tried to get some sleep, he was constantly awakened by his wife "who seemed to cough all night" and by "streetcars pounding along Connecticut Avenue [whose] wheels hadn't been fixed since the Depression." He became irritable and depressed. Three events put him over the edge: After being hit by a car, he was forced to dictate the last chapter of his book from his bed. Then William Keen, his grandfather, died from a stroke. Then his mother died.

Following the tragedies, Freeman believed that he, too, was about to die. Certain he had cancer, he fell into a deep depression. Unable to write, work, or drive a car, he decided to take a cruise and attend a neurology conference in London. The year was 1935—the same year that John Fulton and Carlyle Jacobsen presented their study of Becky and Lucy. After arriving at the conference, Freeman set up a booth describing his work. In the booth next to him was Egas Moniz. The two became friends. Seven months later, when Moniz published the monograph describing his first 20 lobotomy patients, Freeman called it "epoch-making work."

In May 1936, Walter Freeman wrote a letter to Egas Moniz: "I enjoyed particularly your recent work on the reduction of psychiatric symptoms following operation of the frontal lobe. And I am going to recommend a trial of this procedure in certain cases that come under my care." Freeman found out who had made Moniz's lobotomy knives and ordered two of them. They arrived in July 1936. At the time that

Freeman opened the package, hundreds of thousands of patients filled the mental wards of state hospitals across the United States. And the numbers were growing. Something had to be done. Walter Freeman believed he was just the man to do it.

In Moniz's lobotomies, Freeman saw his chance to finally enter the pantheon of medical practitioners. "I recognized that I had done nothing important in either explaining mental disorder or treating it," he had written to Moniz. Lobotomies would soon make Walter Freeman one of the best known physicians in the United States and the world. "It has been said that if we don't think correctly, it is because we haven't 'brains enough,'" he reasoned. "Maybe it will be shown that a mentally ill patient can think more clearly and more constructively with less brain in actual operation."

Walter Freeman's first patient was Alice Hammatt, a 63-year-old housewife from Emporia, Kansas, who complained of "nervousness, insomnia, depression, anxiety, and apprehension" and often "laughed and wept hysterically." According to Freeman, she was vain, afraid of growing old, overly concerned about her thinning hair, rigid, insecure, emotional, fussy, claustrophobic, suicidal, a "master at bitching," and so domineering that her husband led "a dog's life." Alice's husband wanted her to have a lobotomy. She didn't.

On September 14, 1936, Walter Freeman wheeled Alice Hammatt into the operating room. But not before Hammatt had refused to consent to the procedure, concerned that her hair would be cut off. Freeman assured her that her hair would be spared, a clear misrepresentation of what was about to happen. Freeman didn't do the surgery. Like Moniz, he wasn't a surgeon; he was a neurologist. He needed to find a neurosurgeon who was willing to do it. He found him in James Watts, a neurosurgeon at George Washington University Hospital who had received his medical training at the University of Virginia Medical School, and his surgical training at Yale, the University of Chicago, and Breslau, Germany (where he had examined Lenin's brain). Where Freeman was a fast-talking, impatient showman, Watts was slow, gentle, and retiring.

After Alice Hammatt was wheeled into the operating room against her will, Watts shaved her hair, cleaned her scalp with gentian violet, made one-inch-long incisions on both sides of her head, bored holes through her skull with an auger, inserted his lobotomy knife five inches into her brain, and took six cores from each side. The operation took four hours.

Hammatt awoke with a "placid expression," and by evening, "manifested no anxiety or apprehension." When asked about her anxieties, she said, "I seem to have forgotten [them]. It doesn't seem important." Now, however, Hammatt was doing something she had never done before. Holding a paper handkerchief in her hand, she rhythmically rubbed her face and arms as if drying herself. But, at least according to Freeman, she was active and alert, slept well, had a good appetite, and was reading magazines. "I knew as soon as I operated on a mental patient and cut into a physically normal brain, I'd be considered radical by some people," said Freeman, who was nevertheless pleased with the outcome. "We were congratulating ourselves on a brilliant result," he enthused.

Six days after the operation, Alice Hammatt became disoriented, began stuttering, misspelled words, couldn't write legibly or carry on a conversation, and continued with her odd rubbing movements. But she was placid, slept without medication, and lived without a nurse's care. And although her husband and her housekeeper now did most of the work—and although Hammatt was embarrassingly outspoken with friends—her anxieties were gone. Alice Hammatt died five years later of pneumonia. Her husband called her remaining years "the happiest of her life."

Seventeen days after the surgery, Freeman reported the case of Alice Hammatt to the Medical Society of the District of Columbia. "The woman went back home in ten days and is cured," he said. The word "cured" didn't sit well with the audience. Dr. Dexter Bullard, a psychiatrist and superintendent of a private psychiatric hospital in Rockville, Maryland, rose to object. "Walter, you can't say that!" he screamed. Others in the audience nodded in agreement; some shouted their disapproval. A few months after the surgery, Alice Hammatt had a prolonged seizure during which she fell and broke her wrist.

Freeman and Watts published the results of Hammatt's operation in the *Southern Medical Journal.* Titled "Prefrontal Lobotomy in Agitated Depression: Report of a Case," it was the first time the word "lobotomy" had appeared in print. (Moniz and Lima had always used the term "leucotomy.")

In anticipation of the upcoming Southern Medical Association's meeting in Baltimore, Freeman and Watts rushed to perform five more lobotomies. It would be Freeman's second chance to show his American colleagues just how remarkable lobotomies could be. This time, however, Freeman wanted to ensure a better reception. So he called a *Washington Star* reporter named Thomas Henry and offered him an exclusive interview. Days before the meeting, a glowing report of Freeman's work appeared in the *Star.* Henry wrote that lobotomies "probably constitute one of the greatest surgical innovations of this generation . . . It seems unbelievable that uncontrollable sorrow could be changed into normal resignation with an auger and a knife." Before he had even presented his findings, Walter Freeman, at least according to the press, was a hero. "As expected, there was considerable journalistic interest when I arrived in Baltimore," Freeman crowed.

On November 18, 1936, Walter Freeman stood before a group of stunned neurologists and psychiatrists and described the results of his procedure. He explained that all six of his lobotomy patients had improved. No longer did they suffer from disorientation, phobias, confusion, hallucinations, and delusions. And their worry, apprehension, anxiety, sleeplessness, and nervous tension had disappeared. Patients were now calm, content, and much easier to manage. "We are able to say that no patient has died and none has been made worse," said Freeman. "All of our patients have returned home and some of them are no longer in need of nursing care."

Spafford Ackerly of Kentucky rose in support of Freeman's findings. "This is a startling paper," he said. "I believe it will go down in medical history as a noted example of therapeutic courage." But, as had happened when he first presented the case of Alice Hammatt in Washington, D.C., not everyone at the Baltimore meeting was supportive. Joseph Wortis, a Manhattan psychiatrist, argued that lobotomies had merely shocked patients into some degree of normalcy. "I have seen patients get better after a broken leg," said Wortis. Then, Adolf Meyer, the dean of American psychiatry and professor of neurology at the prestigious Johns Hopkins Hospital, rose to speak. "I am not antagonistic of this work, but find it interesting," he said. "There are more possibilities in this operation than appear on the surface." Given his influence, had Meyer been critical, the number of lobotomies performed in the United States might have ended at six. But Meyer had demurred. Encouraged, Freeman and Watts dove back into their work, intent on performing 20 lobotomies by the end of 1936. They wanted more cases to present at a pivotal meeting coming up in Chicago. Adolf Meyer might have felt differently about lobotomies if Walter Freeman had been honest about the outcome of his first six patients. The fifth patient, who had clearly suffered severe, permanent brain damage following inadvertent severing of a cerebral artery, was epileptic and incontinent for the rest of her life.

In February 1937, Walter Freeman stood before hundreds of colleagues at a meeting of the Chicago Neurological Society. It would be his biggest test to date. Freeman and Watts had operated on 20 patients in a period of three months; almost all were women. Freeman remained upbeat, stating that his patients' memories, concentration, judgment, and insight had remained intact and that their ability to enjoy their lives had improved after lobotomies. The only negative, he argued, was that "every patient probably loses something by this operation, some spontaneity, some sparkle, some flavor of the personality." Although Freeman had faced opposition to his procedure in Washington and Baltimore, it was nothing compared to what he was about to encounter in Chicago.

Several doctors argued that the procedure, because it involved blindly removing cores from the brain, was bound to injure cerebral vessels. (In truth, it already had.) Another said that Freeman and Watts couldn't draw any conclusions from their patients because anxiety states could wax and wane; also, patients had been examined for only a short period of time. Another wondered what was to become of the musician or artist whose frontal lobes were mutilated. Others argued that the procedure had no "anatomical basis" and was justified by "loose reasoning." Yet another argued the procedure was "immoral." Freeman countered that "a brain can stand a good deal of manhandling," and that "most of the damage is reversible." Nonetheless, Walter Freeman had been shaken by the criticism; he decided to cancel his next public appearance in St. Louis. "I almost bit the stem of my pipe off trying to regain control of myself," he recalled.

As had been the case with his presentations in Baltimore and Washington, Freeman hadn't been completely honest during the

Chicago meeting. Eight of his original 20 patients had relapsed, requiring repeat lobotomies. Freeman and Watts had been so disappointed with the early outcomes that they'd increased the number of frontal lobe cores from six to nine and drilled deeper into the brain to get them. Two patients had died from cerebral hemorrhages due to the deeper drilling, and another died soon after the surgery from a heart attack. A fourth, who had been a secretary for 13 years, slid into a dysfunctional state and never recovered, spending the rest of her life in a mental facility. Several patients had residual seizure disorders, and some had difficulty moving their arms and legs. William White, the head of St. Elizabeth's Hospital, who was probably best informed about the early lobotomies, refused to allow them to be performed on anyone in his hospital. Like the Portuguese psychiatrists, White had been horrified by what he had seen.

Not all physicians were against lobotomies. The *New England Journal of Medicine,* the nation's foremost clinical journal, wrote that lobotomies were a "rational procedure." And the *New York Times* wrote that the "new operation marked a turning point in treating mental cases." Indeed, on June 7, 1937, a front-page article in the *New York Times* declared—in what read like an advertisement for a patent medicine—that lobotomies could relieve "tension, apprehension, anxiety, depression, insomnia, suicidal ideas, delusions, hallucinations, crying spells, melancholia, obsessions, panic states, disorientation, psychalgia (pain of psychic origin), nervous indigestion, and hysterical paralysis," and that the operation "transforms wild animals into gentle creatures in the course of a few hours."

DESPITE THE TORRENT OF CRITICISM at the Chicago meeting, Walter Freeman and James Watts didn't stay down for long. With *Time, Newsweek,* the *New York Times,* and the *New England Journal of*

*Medicine* on their side, they were back in business, speaking at scientific and medical meetings in New Haven, Boston, New York, Philadelphia, and Memphis, as well as at the prestigious annual meeting of the American Medical Association in Atlantic City. They received hundreds of letters from people across the country begging them to treat mental illness as well as a variety of other medical disorders. One writer asked Freeman to cut out the part of his brain that caused asthma. During the next four decades, more than 20,000 lobotomies would be performed in the United States. Walter Freeman would be personally responsible for almost 4,000 of them.

TODAY WE VIEW LOBOTOMIES AS cruel, freakish, and comical. They're a drink (the "Frontal Lobotomy" is made with amaretto, Chambord, and pineapple juice), a saying (Tom Waits's, "I'd rather have a bottle in front of me than a frontal lobotomy"), and a slogan (during the Iraq War, protesters wore T-shirts with a picture of George W. Bush above the words, "Ask me about my lobotomy"). Lobotomies now share a shelf in the dusty cabinet of medical sideshows next to whips, chains, snake pits, truth serums, phrenology machines, and trephining, the ancient ritual of drilling holes in the brain to loose the evil spirits. So why were lobotomies so readily accepted, indeed sought after, in the late 1930s through the early 1970s? Three reasons.

First, psychiatrists, families, and patients were desperate to do something, anything to treat untreatable mental disorders, most commonly schizophrenia. And there weren't any other good options.

Second, state mental hospitals were bursting at the seams. The number of patients rose from 159,000 in 1909 to 480,000 in 1940, a rate twice the growth rate of the general population. Indeed, in the 1940s and 1950s, as many people were hospitalized with mental disorders in the United States as for all other diseases combined. Lobotomies provided a seductive way out of an unmanageable situation.

Third, conditions in state hospitals were hideous. In May 1946, an article written in *Life* magazine titled "Bedlam 1946" described just how badly things had deteriorated. Patients were beaten, abandoned, provided little clothing, put in dark, damp, padded dungeons, restrained in straitjackets for weeks, and forced to lie in their own excrement. Facilities resembled "concentration camps on the Belsen pattern." The staff—which was often composed of uneducated transients who were plucked from local jails—raped, sodomized, and occasionally murdered patients. And physicians were nowhere to be found; the ratio of patients to physicians was 250 to 1.

Another reason that lobotomies were not only accepted but embraced can be found in a comment made by Joseph Wortis, the Manhattan psychiatrist who had argued during the Baltimore meeting that Freeman's lobotomies had merely "shocked" patients into normalcy. In 1940s America, the key to many psychiatric therapies was to shock patients out of their illnesses. Lobotomies weren't much worse than what psychiatrists had already been doing.

IT STARTED IN THE MIDDLE AGES. To shock patients with mental illness, physicians either nearly drowned them or forced them to walk down a dark hallway at the end of which was a snake pit. In the early 20th century, four other therapies were based on the same concept, all of which were, according to one observer, "like trying to fix a watch with a hammer."

In 1917, Julius Wagner von Jauregg invented malaria therapy. Von Jauregg found that patients suffering from paralysis and mental illness caused by syphilis could be treated by injecting them with blood from patients with malaria. The goal was to induce fevers as high as 106°F, which were felt to be curative. For his discovery, von Jauregg won the Nobel Prize in medicine in 1927. It was the first Nobel Prize ever awarded for a psychiatric therapy. Egas Moniz, for his invention of lobotomies, won the second. At this point, it seems reasonable to wonder whether Nobel Prizes awarded in the first half of the 20th century came in Cracker Jack boxes. But the truth is that von Jauregg's malaria therapy really did work on patients suffering from syphilis, which is caused by a spirochete. As it turns out, spirochetes are sensitive to high temperatures. After patients improved, they were given quinine to end the malaria infection. The problem with malaria therapy was that it was also used for many other psychiatric illnesses, for which it offered nothing. (Remarkably, malaria therapy isn't dead. Believing that they are suffering from "chronic Lyme disease," some Americans still travel to Mexico to get injected with malaria parasites.)

In 1930, Manfred Sakel invented insulin shock therapy. Sakel, who worked in Vienna, had accidentally given too much insulin to a morphine addict and found that the mistake cured the addiction. He then tried insulin therapy on 15 more patients, all, according to him, with the same result. Then he tried his therapy on people with schizophrenia, claiming an 88 percent cure rate. Following Sakel's lead, patients in the United States were given larger and larger doses of insulin until their blood sugars were so low that they slipped into a coma. Clinicians would then administer varying amounts of glucose by nasogastric tube, hoping to maintain the coma without killing the patient. Typically, patients were in a coma for one to two months. Many died.

In 1935, Ladislas Joseph Meduna, a Hungarian researcher, invented Metrazol shock therapy. Metrazol caused seizures, and Meduna believed that seizures could treat schizophrenia. He claimed that after treating a patient with catatonic schizophrenia who had been lying in bed for four years, the man got up, dressed himself, put on his hat, and walked out of the hospital. Meduna treated ten more patients, supposedly with the same result.

In 1938, Ugo Cerletti, working in Italy, invented the shock therapy to end all shock therapies: electroshock. Cerletti first tried his therapy on a man with schizophrenia who had been wandering around a police station. He attached electrodes to each side of the man's head and flipped the switch. The man stopped breathing, turned blue, suffered a seemingly endless stream of seizures, and recovered. Cerletti insisted that from that point on, the man acted normally. Electroshock therapy was the easiest and most commonly used of the shock therapies.

By 1942, at least 75,000 psychiatric patients, mostly with schizophrenia, had received some form of shock therapy. Today, electroshock, which is used for patients with severe depression, is the only one that has survived. When lobotomies were first introduced in the United States, they were competing with what one psychiatrist called "the therapies of despair."

Between 1936—when Walter Freeman and James Watts performed their first lobotomy—and 1942, about 300 lobotomies were performed in the United States. In 1943, another 300 were performed; by 1947, another thousand; in 1948, another 2,000; and in 1949, another 5,000. By August 1949, more than 10,000 lobotomies had been performed. About 60 percent were performed in state mental hospitals, mostly on women, even though fewer women were housed in state institutions than men. By the end of 1951, Freeman and Watts and their trainees had performed more than 18,000 lobotomies. People were lining up to get them. Middle-aged women wanted lobotomies to cure their depression; college students wanted them to cure their neuroses; and parents wanted them to cure their misbehaving children.

Although lobotomies had become enormously popular, the reason that the number increased so quickly and so dramatically was that Walter Freeman had changed the way that he did them—creating what could reasonably have been called the "drive-through lobotomy."

In January 1946, Freeman performed a lobotomy on Sallie Ionesco. This time, however, James Watts didn't do the surgery; Freeman did. Freeman didn't do it in an operating room; he did it in his office. He didn't sterilize his instruments or sterilize the operative site. He didn't use general anesthesia; rather, he used electroshock to anesthetize the patient (even though electroshock caused unconsciousness for only a few minutes). And he didn't use a scalpel to cut open the skin at the side of the skull or an auger to bore holes; he used an ice pick bearing the label "Uline Ice Company" that he had taken from his kitchen drawer. Freeman inserted the ice pick into the bone on the upper and inner aspect of Sallie Ionesco's eye socket, drove it about three inches into her brain with a small hammer, and wiggled it around. Then he repeated the procedure on the other eye socket. Freeman's new ice pick lobotomy didn't take four hours; it took seven minutes. And, at least according to Freeman, anyone could do it—even if they hadn't received any formal surgical training. James Watts knew nothing about this. When he walked into Walter Freeman's office and saw an ice pick sticking out of Sallie Ionesco's face, he was appalled. Watts believed that any procedure that required disruption of brain tissue should be performed in an operating theater where the brain could be visualized. Otherwise, the lobotomist increased the chances of accidentally tearing a cerebral artery, causing fatal hemorrhage. Watts and Freeman never worked together again.

With his new ice pick method, Walter Freeman moved lobotomies into the express lane of quick-fix cures. He hopped in his car—ice picks in his jacket pocket—and sped around the country demonstrating his procedure to anyone who would listen. He visited state mental hospitals in California, Texas, Arkansas, Minnesota, Ohio, New York, Washington, Missouri, and Maryland, logging more than 86,000 miles. (He called his car the "Lobotomobile.") At the Weston State Hospital in West Virginia, Freeman operated on 228 patients in 12 days; performing 22 operations in 135 minutes—an average of 6 minutes per patient. When he was finished, Freeman had visited 55 hospitals in 23 states as well as psychiatric institutions in Canada, the Caribbean, and South America. His daughter called him "The Henry Ford of Psychiatry."

Ever the showman, Freeman became so comfortable with the procedure that he could disrupt the frontal lobes behind each eye at the same time. A nursing student named Patricia Derian witnessed Freeman conducting one such operation in 1948. "He looked up at us, smiling," she recalled. "I thought I was seeing a circus act. He moved both hands back and forth in unison, cutting the brain identically behind each eye. It astonished me that he was so gay, so high, so 'up.'" (Today, radiologists reading MRIs and CT scans are occasionally surprised to find the characteristic tracts of disrupted brain in patients who had previously been subjected to Walter Freeman's ice pick lobotomies.)

IN 1950, Freeman wrote a book titled *Psychosurgery*. In it, he reported the outcomes of several hundred lobotomies. Freeman concluded that his invention had not only relieved America of its burden in state psychiatric facilities, it had also cured many other citizens of their anxieties and neuroses. A closer look at Freeman's book, however, shows just how low he had set his bar for success.

During the first few weeks after their lobotomies, almost all patients suffered similar symptoms. They would lie in their beds like "wax dummies" and have to be turned constantly by visiting nurses or family members to keep from getting bedsores. All were profoundly indifferent to their surroundings. They didn't seem to care about anything. Worse, they had lost any sense of decorum. One well-bred woman defecated in a wastebasket, thinking it was a toilet. Other patients would "vomit into their soup plates and start eating out of the plate again before the nurse [could] take it away." But to Freeman the bottom line was that virtually all of these patients were less disruptive on the wards, much easier to handle. He had created, in his own words, a "surgically induced childhood." About 25 percent of lobotomy patients never progressed beyond this stage and remained institutionalized. Some became disruptive again and were forced to have second and even third lobotomies.

Most lobotomy patients were able to leave the wards and return to their families. However, all lacked energy, were lethargic, and, according to Freeman had "lost interest in themselves." Families had to dress and undress the patient. And patients typically lost any sense of shame. They would present themselves naked to strangers, take food off of other people's plates at the dinner table, and spend hours "like little children" in the bathtub, "squirting water around." Patients became outspoken, unrestrained, and silly; nothing was taken seriously. They would get lost while traveling. One woman, who used to read Victorian novels, still read them, but could no longer remember what they were about. Freeman described these patients as "adjusting to the level of a domestic invalid or household pet." Still, he reasoned that lobotomies had virtually eliminated their anxieties and neurosis.

Some lobotomy patients were able to leave home and go to work, but never at the level of performance seen before the operation.

Professors waited on tables. Cashiers couldn't keep numbers straight. Saleswomen could no longer give correct change. Musicians could still play, but their music had become mechanical and heartless. Most of those who tried to go back to work were fired.

Side effects from the procedure were also grim. As James Watts had feared, three of every hundred people died from their ice pick lobotomies, always from uncontrolled bleeding after tearing a cerebral artery. Another three of every hundred suffered permanent, unremitting seizures. Many others were unable to control their bowels or bladder. Perhaps most outrageous, Freeman had operated on children, 11 of whom were described in his book. One was only four years old. Two of the 11 died from cerebral hemorrhages.

As the number of lobotomies increased, many physicians became critical, calling the procedure "medical sadism," "mutilation," and "partial euthanasia," arguing that Freeman had merely replaced one awful disorder with another. When John Fulton—the Yale researcher who had inspired Egas Moniz to perform the first human lobotomy—heard about Freeman's ice pick lobotomies, he said, "What are these terrible things I hear about you doing with lobotomies in your office with an ice pick? I have just been to California and Minnesota and heard about it in both places. Why not use a shotgun? It would be quicker!"

Despite these protests, no professional, medical, or ethical society, including the American Medical Association and the American Psychiatric Association, stood up to oppose lobotomies. Worse, the media continued to misinform the public. Headlines like "Surgeon's Knife Restores Sanity to Nerve Victims," "No Worse Than Removing a Tooth," and "Wizardry of Surgery Restores Sanity to Fifty Raving Maniacs" routinely appeared in newspapers across the country. One cartoon in the *American Weekly* published in 1946 told the story of a "shy, mousy little bookkeeper, the kind that is the butt of all office pranksters," who, following a lobotomy, became a "gregarious hail-fellow-well-met type, who could sell anything to anybody"—a captain of industry.

In 1941, one of the most influential articles on the procedure, titled "Turning the Mind Inside Out," appeared in the *Saturday Evening Post.* The article was written by Waldemar Kaempffert, who was also an editor at the *New York Times:* "There must be at least two hundred men and women in the United States who have had worries, persecution complexes, suicidal intentions, obsessions, indecisiveness, and nervous tensions literally cut out of their minds with a knife by a new operation on the brain." Walter Freeman later said that without the press, lobotomies would never have been so widely accepted.

When Walter Freeman invented the ice pick lobotomy, his goal was to relieve state institutions of the financial burden of caring for the indigent. But Freeman also lobotomized the rich and famous. Tennessee Williams's sister, Rose, the subject of his plays *The Glass Menagerie* and *Suddenly Last Summer,* was lobotomized. As was beat poet Allen Ginsberg's mother. But Freeman's most famous victim was Rosemary Kennedy, the daughter of Joseph and Rose Kennedy, and the sister of President John F. Kennedy, Attorney General Robert F. Kennedy, and Senator Ted Kennedy.

Rosemary Kennedy suffered from mild developmental delay, quickly falling behind her eight brothers and sisters. In a family that took pride in its physical and intellectual prowess, Rosemary was an embarrassment. But she was certainly functional. Before her lobotomy, Rosemary had traveled overseas without a chaperone, participated in sailboat races, and, although it was a struggle, learned to read and write. When she was 15 years old, she sent a letter to her father:

"I would do anything to make you so happy," she wrote. As she got a little older, however, she would occasionally suffer from mild outbursts of anger, raising her voice and flailing her arms. Today, Rosemary Kennedy would likely be diagnosed with mild developmental delay and treated with occupational and behavioral therapies. But to Joseph Kennedy, his daughter's behavior was intolerable—a threat to his and his other children's political futures. He wanted Rosemary fixed. So he took her to a prominent Boston neurologist, asking about this new surgery. The neurologist advised against it. Mild mental retardation, he argued, was not a reason to get a lobotomy. Then Kennedy sought out Walter Freeman, who diagnosed Rosemary with "agitated depression." Freeman was convinced he could treat her. Joseph Kennedy never told his wife about his plan.

In November 1941, Walter Freeman lobotomized Rosemary Kennedy. She was only 23 years old. Jackson cut two holes on the sides of Rosemary's head, inserted his lobotomy knife, and asked her to count, sing songs, tell stories, and name the months of the year to judge how many cores to remove from the brain. This way, Freeman could determine whether her intellectual functions were still intact. But Freeman went too far. After he removed his fourth and final core, Rosemary emerged from the surgery physically and mentally disabled. The nurse who had attended the operation was so upset that she quit the profession.

Joseph Kennedy, however, remained optimistic. He had his daughter admitted to the Craig House, a private psychiatric hospital in Beacon, New York. Situated on 380 acres of rolling countryside, the Craig House had an indoor swimming pool, a golf course, stables, an arts and crafts center, and a highly trained medical staff. Care at the Craig House cost the equivalent today of $250,000 a year. After several months of intense therapy, Rosemary regained her ability to walk. But she could no longer read, spoke only a few words, was unable to take care of herself, and had lost all memory of her friends and family.

Rose Kennedy, who periodically sent out long letters to her children updating the family's activities, never mentioned Rosemary again. And she didn't visit her daughter for 20 years. Joseph Kennedy, who later transferred Rosemary to the St. Coletta School for Exceptional Children in Jefferson, Wisconsin, didn't visit Rosemary for the last 25 years of his life. Indeed, the only person to visit Rosemary before she died in 2005 was her brother John, who privately visited her during a campaign stop in Wisconsin in 1958. Rose Kennedy later wrote, "Rosemary's was the first of the tragedies that were to befall us."

By the middle of the 20th century, lobotomies had become such an integral part of American culture that they found their way into books like Robert Penn Warren's *All the King's Men* (1946), plays like Tennessee Williams's *Suddenly Last Summer* (1958), movies like *A Fine Madness* (1966), *Planet of the Apes* (1968), *A Clockwork Orange* (1971), *Terminal Man* (1974), *One Flew Over the Cuckoo's Nest* (1975), *Frances* (1982), *Repo Man* (1984), *A Hole in One* (2004), and *Asylum* (2008), and songs like the Ramones' "Teenage Lobotomy" (1977).

The gradual change in American attitudes toward lobotomies was evident in these works. In the 1959 movie *Suddenly Last Summer,* which starred Elizabeth Taylor and Katharine Hepburn, the lobotomist was played sympathetically by Montgomery Clift. By the 1970s, however, the attitude toward lobotomies had changed. Now they were portrayed as a means to punish nonconformity. For example, in the 1975 movie *One Flew Over the Cuckoo's Nest,* based on the novel by Ken Kesey, the main character, Randle Patrick McMurphy, played by Jack Nicholson, fakes a mental illness to escape a rape conviction. In the ward, he rallies a group of submissive men to rise up against the hospital matriarch, Nurse Ratched, who emasculates them with drugs and faux therapy sessions. During the final confrontation, when McMurphy tries to strangle Nurse Ratched, he is lobotomized into submission: a tragic figure.

By the early 2000s, lobotomies were no longer seen as a tool for institutional control, but rather as a subject of horror movies. In *Asylum* (2008), six freshmen college students discover that their newly renovated dorms were originally a private mental hospital. Flashbacks include a seizing boy strapped to a bed, a girl wearing a straitjacket made of barbed wire, and a boy with an ice pick sticking out of his eye socket, lorded over by a tall man with a small hammer. The last image was perhaps the most frightening. Not because it was the most gruesome, but because it wasn't fictional. The boy was Howard Dully, who later wrote a book about his experiences, and the lobotomist was Walter Freeman. Movies like *Asylum* show how, according to one critic, "it requires only a small shift in our credibility for the surgeon to become the slasher."

Although lobotomies were eventually reviled, nothing hastened their demise more than psychoactive drugs. In 1954, the Food and Drug Administration licensed Smith, Kline & French's chlorpromazine, the first drug offered for the treatment of schizophrenia. The drug was called Thorazine, in honor of Thor, the hammer-wielding Norse god of thunder responsible for protecting humankind. Thorazine worked, clearly reducing the hallucinations and delusions of people with schizophrenia (something that lobotomies never did). By 1955, Tofranil, an antidepressant, and Miltown, an antianxiety drug, had also become available. Now psychiatrists could prescribe mind-numbing drugs whose side effects could be reversed by simply stopping the medication—instead of mind-destroying surgeries, whose side effects were irreversible.

WALTER FREEMAN's career didn't end well.

In 1954, the same year that Thorazine was introduced, Freeman decided to leave George Washington University and move to northern California, where a friend had offered him a position in the Palo Alto Clinic. Freeman asked the medical faculty of George Washington University to make him a professor emeritus, which it refused, offering only that he could take a one-year leave of absence at the end of which he could resign.

When Freeman arrived in California, he found that other doctors in the Palo Alto Clinic had changed their minds, finding him too controversial. Still, there were plenty of other clinics and hospitals in the area where he could perform his lobotomies. In 1961, in front of a large group of neurologists and psychiatrists at the Langley Porter Clinic, Walter Freeman presented three of the seven adolescents he had lobotomized between 1958 and 1960. He wanted those in attendance to see how well his patients performed. Like a ringmaster in a circus of the absurd, Freeman urged the children to demonstrate their mental and physical skills. When one boy was slow to respond, Freeman raised his voice. The child, clearly frustrated, screamed, "I'm doing the best I can!" This exchange didn't exactly win the crowd's sympathy; Freeman was practically booed off the stage. He responded angrily, taking a box of Christmas cards that he had received from his patients and throwing them onto a table. "How many Christmas cards did you get from your patients!?" he yelled. No longer a respected neurologist, Walter Freeman had become a pariah. Freeman still traveled across the country doing lobotomies and training others to do them, but he was clearly slowing down, performing only eight lobotomies in 1965.

In 1967, when he was 72 years old, Walter Freeman performed his last lobotomy. When the woman died of cerebral hemorrhage, he lost his license to practice medicine.

In the final pages of his autobiography, Walter Freeman hoped that one day the hospitals of West Virginia would stand as a "monument to the success of lobotomy." He picked West Virginia because of the large number of lobotomies he had performed at Weston State Hospital in the early 1950s. In a sense, Freeman has gotten what he wanted.

In 2008, Weston State Hospital reopened as a tourist attraction under the name Trans-Allegheny Lunatic Asylum. The asylum has now been featured on Syfy Channel's *Ghost Hunters* and the Travel Channel's *Ghost Adventures*. Apparently, it's haunted. And the ghost doing the haunting is Walter Freeman. At the entrance to the asylum is a picture of Freeman and a plaque describing his work: "Freeman was discredited by the medical profession and lost his license to practice in hospitals. He spent the remainder of his life seeking redemption, pathetically traveling around the country trying to locate his former patients to prove he improved their lives. He died of colon cancer in 1972 at the age of 77. Today Freeman is regarded by many as a monster, an American Mengele."

Not exactly the monument Walter Freeman had been hoping for.

The lesson from lobotomies is far easier to make than to follow: **Beware the quick fix.**

When Joseph P. Kennedy wanted to cure his daughter's mild developmental delay, he sought the counsel of Walter Freeman, who told Kennedy that a simple fix was at hand. The procedure would take only a few minutes, wouldn't require general anesthesia, and would restore his daughter to a level of emotional maturity similar to her brothers and sisters. All the money and time Kennedy had spent on tutors and private schools trying to get Rosemary up to speed hadn't been necessary. All she really needed was one carefully placed surgical knife in her brain. Even though a noted surgeon had recommended Rosemary's lobotomy, the reason that it sounded too good to be true was that it was too good to be true. A healthy skepticism by Kennedy could have spared his daughter a life of severe physical and mental disability. All Kennedy had to do was ask to see others who had been lobotomized. But he didn't want to see them. He wanted to believe in the magic. And for that, his daughter paid an enormous price. As did the thousands of other patients and their families who wanted to believe that complex psychiatric disorders could be treated with a five-minute surgery.

As we'll see in the final chapter, when it comes to children with autism, the lesson of the quick fix remains unlearned.

# Chapter 6

# THE MOSQUITO LIBERATION FRONT

*Hey farmer, farmer*
*Put away that DDT now*
*Give me spots on my apples*
*But leave me the birds and the bees*
*Please!*

—Joni Mitchell, "Big Yellow Taxi"

Olga Huckins was angry. On January 29, 1958, she wrote a letter to the *Boston Herald.*

The previous summer—in an effort to kill growing numbers of gypsy moths, tent caterpillars, and mosquitoes—state authorities had sprayed DDT over large areas of Pennsylvania, New York, and New England. Huckins, who lived near a bird sanctuary in Duxbury, Massachusetts, was horrified by what happened next. "We picked up three dead [robins] the next morning," she wrote. "They were birds that had lived close to us, trusted us, and built their nests in our trees year after year . . . All these birds died horribly and in the same way. Their bills were gaping open, and their splayed claws were drawn up to their breasts in agony."

Huckins wasn't alone in her anger. Many residents had written letters, sickened by the aftermath of the spraying. Health officials were unbowed. But Olga Huckins refused to be ignored. She sent a copy of her *Boston Herald* letter to her friend, Rachel Carson. Four years later, Carson published a book about it. Called *Silent Spring*, it became an international best seller, alerting the world to the dangers of pesticides, landing Carson on national television programs and in front of congressional hearings, winning praise from people as diverse as President John F. Kennedy, Supreme Court Justice William O. Douglas, and singer-songwriter Joni Mitchell, and making Carson one of the most famous and most influential women in the United States. Unfortunately, Rachel Carson had made one tragic mistake.

RACHEL LOUISE CARSON was born on May 27, 1907, in Springdale, Pennsylvania, the third of three children. Her father, Robert, was variously an electrician, an insurance salesman, and a night watchman. Her mother, Maria, gave up teaching to raise the children.

Although Springdale was a hardscrabble town known for its glue factory, drab streets, and blue-collar workers. Maria filled Rachel's childhood with the wonders of nature. Walking hand in hand through the nearby woods, orchards, and fields, or strolling along the banks of the Allegheny River, Maria described in vivid detail the wealth of life around them. So devoted was Maria to her youngest child that, with the exception of a few years in college, she never left her side. More than anything else, Maria encouraged Rachel to write.

In 1922, when she was 15 years old, Rachel wrote an article for *St. Nicholas* magazine that offered a glimpse into her future. Accompanied by her dog, Pal, Carson described woods where the "majestic silence [was] interrupted only by the rustling breeze, and the cheery, 'witch, witchery' of the Maryland yellow throat." She lovingly described the music of orioles, bobwhites, cuckoos, and hummingbirds, and a nest "containing four jewel-like eggs." Carson had found a utopia far from the grimy streets of Springdale. Another world. An Eden. A place where she could immerse herself in the intricacies of nature. A place where the man-made stench of Springdale's glue factory had faded into the distance.

After graduating from high school in 1925, Carson attended the Pennsylvania College for Women (now Chatham College) in Pittsburgh. Maria visited her almost every weekend. Although Carson entered as an English major, she fell in love with science, taking courses in botany, zoology, histology, microbiology, and embryology. In 1929, she graduated magna cum laude. That summer, Rachel earned a scholarship to the prestigious Marine Biological Laboratory at Woods Hole, Massachusetts. Here, she fell in love with the sea. Encountering young mullet fish at night, she wrote, "I stood knee deep in that racing water, and could barely see those darting silver bits of life for my tears. That was when I first began to let my imagination go down under the water."

In the fall of 1929, Carson attended Johns Hopkins University in Baltimore, determined to earn a Ph.D. degree. It wasn't to be. Although Rachel submitted a thesis for her master's degree in 1932—about kidney development in catfish—her professors didn't believe she had what it took to be a scientist. Abandoned by her mentors, she never performed another scientific experiment—and never received her Ph.D.

During the next few years, Carson contributed articles to the *Baltimore Sun* and the *Richmond Times* about tuna fishing off Nova Scotia, oyster farming in the Chesapeake Bay, starlings overwintering in Baltimore, and eels migrating to the Sargasso Sea. She also wrote about how certain species were becoming dangerously depleted, like elk, heath hens, salmon, shad, canvasback ducks, pronghorn antelope, mountain goats, moose, and bears. Carson's writing had taken a darker turn, now focusing almost exclusively on scarcity and extinction.

In 1935, Rachel Carson dropped out of Johns Hopkins to work as a field aide for the U.S. Bureau of Fisheries in College Park, Maryland. Her job was to write pamphlets and press releases. The following year, she was appointed to a full-time position at the bureau as a junior aquatic biologist. There, she wrote short scripts for a radio program called "Romance Under the Waters." Two years later, she burst onto the American scene.

In September 1937, Carson published an article in the *Atlantic Monthly* titled "Undersea." "The conquest of Mt. Everest has passed into history," she wrote. "But although the flags of explorers have waved on the highest peaks of the world and fluttered on the frozen rims of the continents, a vast unknown remains: the world of waters." Inspired by Jules Verne's *Twenty Thousand Leagues Under the Sea,* she continued: "The ocean is a place of paradoxes. It is the home of the great white shark, 2,000-pound killer of the seas, and the 100-foot blue whale, the largest animal that ever lived. It is also home of living things so small that your two hands might scoop up as many of them as there are stars in the Milky Way." Carson signed it "R. L. Carson," certain that no one would read a scientific article if they knew a woman had written it.

Editors from Simon & Schuster read "Undersea" and loved it. They wanted Carson to write a book. On November 1, 1941, Rachel Carson published *Under the Sea-Wind,* this time using her full name. Although written for adults, the book had a childlike sense of wonder. *Under the Sea-Wind* told the story of Silverbar, a sanderling that migrated from the Arctic Circle to Argentina; Scomber, a mackerel that traveled from New England to the continental shelf; and Anguilla, an American eel that journeyed to the Sargasso Sea, joining thousands of other eels that had come to spawn. "There is poetry here," said one reviewer.

More reviews followed. The *New York Times* called it "a beautiful and unusual book; a breathtaking canvas of the fierce struggle for life." The *New Yorker, Christian Science Monitor, New York Herald Tribune,* and *New York Times Book Review* also sang its praises, unequivocally. But the timing wasn't right. One month later, the Japanese attacked Pearl Harbor and Americans turned their attention to war. By June 1942, fewer than 1,200 books had been sold; Carson's royalties amounted to $689.17. "The world received the book with superb indifference," she lamented.

Carson didn't blame World War II for poor sales. She blamed her publicists at Simon & Schuster, whom she believed had failed to adequately promote her book. She asked out of her contract, never publishing with them again. Despite this personal setback, Carson's stock at the Bureau of Fisheries, which was now called the Fisheries and Wildlife Service, continued to rise. By 1949 she was editor in chief of all scientific publications.

On July 2, 1951, Oxford University Press published Carson's second book, *The Sea Around Us*. One month before publication, excerpts were printed in the *New Yorker*. The response was overwhelming; more letters were written to the magazine than at any time in its history. Carson became an instant celebrity. Even Walter Winchell, the acerbic radio pundit, commented on how much he was looking forward to reading Carson's new book.

The opening of *The Sea Around Us* read like the beginning of Genesis: "Imagine a whole continent of naked rock, across which no covering mantle of green had been drawn . . . Imagine a land of stone, a silent land, except for the sounds of the rains and winds that swept across it. For there was no living voice and nothing moved over its surface except the shadows of the clouds."

The *New York Herald Tribune* called *The Sea Around Us* "one of the most beautiful books of our time." Teddy Roosevelt's daughter said it was the best thing she'd ever read. Three weeks after publication, *The Sea Around Us* was #5 on the *New York Times* best-seller list behind only Thor Heyerdahl's *Kon-Tiki,* Herman Wouk's *Caine Mutiny,* James Jones's *From Here to Eternity,* and J. D. Salinger's *The Catcher in the Rye*. In September, *The Sea Around Us* was #1, where it stayed for 39 weeks—a record. By November, 100,000 copies had been sold, by March, 200,000. The Book-of-the-Month Club selected it and *Reader's Digest* condensed it. Four thousand copies were selling every week. When the dust settled, *The Sea Around Us* had sold more than 1.3 million copies and had been translated into 32 languages.

Then Hollywood stepped in, making *The Sea Around Us* into a movie. Directed by the "master of disaster," Irwin Allen—best known for movies and TV series such as *The Towering Inferno, The Poseidon Adventure, The Swarm, Voyage to the Bottom of the Sea,* and *Lost in Space*—*The Sea Around Us* won an Academy Award for Best Documentary Feature. Carson hated the movie. And hated all the tinsel and cardboard publicity that came with it. When Jacques Cousteau offered to take her on a voyage on his *Calypso,* she refused.

Despite Carson's disdain for the spotlight, the awards and recognitions kept coming, including the National Book Award, arguably the single most coveted book prize in the United States. Editors of the nation's leading newspapers voted Carson "Woman of the Year."

Fortune followed fame. Before *The Sea Around Us* was published, Oxford University Press gave Carson a $1,000 advance. Later, she collected $7,200 from the *New Yorker* for its serialization, $10,000 from *Reader's Digest* for its condensation, $20,000 in royalties for continued sales, and another $20,000 from RKO for the movie rights, all of which added up to more than four times her annual salary. Financially secure, Carson quit her job at the Fisheries and Wildlife Service to devote herself full-time to writing. Despite the enormous financial success of *The Sea Around Us,* Carson still wasn't satisfied, believing that Oxford—which had taken out full-page ads in the *New York Times, New York Herald Tribune,* and *Chicago Tribune*—hadn't done enough to promote her book.

Rachel Carson's fame, sadly, would be short-lived. During the writing of *The Sea Around Us,* when she was 43 years old, Carson had two small lumps removed from her breast. The pathologist labeled them benign—a diagnosis that would later be called into question.

IN OCTOBER 1955, Rachel Carson published her next book, *The Edge of the Sea,* a tour guide for the casual adventurer. Again the *New Yorker* serialized it; the critics praised it, ("Carson has done it again in this wise and wonderful book"); the public bought it (more than 70,000 copies sold as it rocketed to #4 on the *New York Times* best-seller list); and Carson lamented it, claiming that Houghton Mifflin—the next up in her string of publishers—hadn't adequately promoted it even though they had spent more than $20,000 on advertising.

Although she didn't have a Ph.D. or an institutional affiliation, by the early 1960s, just before she published her next book, Rachel Carson was America's most famous science writer. The public loved her, the media trusted her, and the government turned to her.

In 1962, Rachel Carson published *Silent Spring,* an angry, raging, no-holds-barred polemic against pesticides—especially one called DDT. From the first page—a quote from E. B. White—*Silent Spring* made it clear that this was not a subject for equivocation. "I am pessimistic about the human race," wrote White, "because it is too ingenious for its own good. Our approach to nature is to beat it into submission. We would stand a better chance of survival if we accommodated ourselves to this planet and viewed it appreciatively instead of skeptically and dictatorially."

The first chapter, titled "A Fable for Tomorrow," began innocently: "There was once a town in the heart of America where all life seemed to live in harmony with its surroundings." A town with "prosperous farms, with fields of grain, and hillsides of orchards." A town where "white clouds of bloom drifted above the green fields." A town "famous for the abundance and variety of its bird life." But a dark cloud was hovering in the distance. "Then a strange blight crept over the area and everything began to change . . . Mysterious maladies swept the flocks of chickens . . . the cattle and sheep sickened and died . . . roadsides . . . were now lined with browned and withered vegetation as though swept by fire . . . streams were lifeless . . . everywhere there was the shadow of death." Birds, especially, had fallen victim to this strange evil. "The birds . . . where had they gone? Feeding stations in the backyard were deserted [and] the few birds seen anywhere were moribund; they trembled violently and could not fly." In a town that had once "throbbed with the dawn chorus of robins, catbirds, doves, jays, wrens, and scores of other bird voices there was now no sound, only silence." A silent spring. Birds weren't alone in their suffering. According to Carson, an increasing number of children were suffering from birth defects, liver disease, and leukemia. And women were suffering from infertility. "There had been several sudden and unexplained deaths," wrote Carson, "not only among adults but even among children, who would be stricken suddenly while at play and die within a few hours."

Carson had made it clear from the start that she wasn't talking about something that *might* happen—she was talking about something that *had* happened. "Many real communities have already suffered," she wrote. "A grim specter has crept upon us almost unnoticed." *Silent Spring* was a book about pesticides—a book that read like stories from the Brothers Grimm. Worse, despite all our efforts, the insects were fighting back, now stronger and more voracious than ever. Our response, it seemed, was simply to make more chemicals; between 1947 and 1960, production of synthetic pesticides increased from 124 million pounds to 638 million pounds. According to Rachel Carson, our war against nature had become a war against ourselves.

Although the U.S. Department of Agriculture had placed some restrictions on DDT a few years before the publication of *Silent Spring* (because of stream pollution), Carson's book ignited a movement that would eventually eliminate the pesticide from the face of the earth.

On August 29, 1962, one month before *Silent Spring* was published, President John F. Kennedy appeared at a press conference. One reporter asked, "Mr. President, there appears to be a growing concern among scientists as to the possibility of dangerous long-range side effects from the widespread use of DDT and other pesticides. Have you considered asking the Department of Agriculture or the Public Health Service to take a closer look at this?" "Yes," replied Kennedy. "And I know that they already are. I think, particularly, of course, since Miss Carson's book, they are examining the issue." Carson's book was already having an impact. And it hadn't even been published yet. Kennedy learned about it from three excerpts that had been serialized in the *New Yorker* that summer.

President Kennedy wasn't the only one who had taken notice. Just before publication, scores of newspapers and magazines had reviewed the book, virtually all favorably. Walter Sullivan, a science reporter and editor wrote, "In her new book, [Rachel Carson] tries to scare the living daylights out of us and, in large measure, succeeds."

Two weeks after publication, *Silent Spring* sold 65,000 copies.

Two weeks after that, in October 1962, the Book of the Month Club sold 150,000 more copies, helped in no small part by an endorsement from U.S. Supreme Court Justice William O. Douglas, who called it "the most important chronicle of this century for the human race."

By Christmas, *Silent Spring* was #1 on the *New York Times* best-seller list, where it remained for 31 weeks. Sales weren't limited to the United States. Translated into 22 languages, *Silent Spring* became an international best seller, described as "one of the most influential books in the modern world." Later, the *New York Times* and New York Public Library listed *Silent Spring* as one of the 100 most important books of the 20th century.

Carson had become an environmental guru. One month after publication of *Silent Spring,* a journalist named Jane Howard profiled Carson in a *Life* magazine article titled "The Gentle Storm Center: A Calm Appraisal of *Silent Spring,*" Howard called Carson "a formidable adversary" and painted her as the leader of a powerful new movement. Carson didn't see it that way. "I have no wish to start a Carrie Nation crusade," said Carson, referring to a prominent leader in America's temperance movement. "I wrote the book because I think there is a great danger that the next generation will have no chance to know nature as we do—if we don't preserve it, the damage will be irreversible." Howard also tried to portray Carson as an early feminist. Again, Carson resisted. "I'm not interested in things done by women or men," she said, "but in things done by people."

Two months after publication, Eric Sevareid interviewed Rachel Carson for his popular television program *CBS Reports,* the precursor to *60 Minutes*. For two days in November 1962, Sevareid talked with Carson, who was now suffering from metastatic breast cancer, a disease that would take her life 17 months later. Thin, haggard, and wearing a heavy black wig to hide the hair loss she had suffered following radiation therapy, Carson marshaled on. But her illness was evident. At the end of the interview, knowing that it would be months before the show aired, Sevareid turned to his producer, Jay McMullen, and said, "Jay, you've got a dead leading lady."

On April 3, 1963, one year before Rachel Carson died from metastatic breast cancer, *CBS Reports* aired "The Silent Spring of Rachel Carson." The show pitted Carson against an established scientist named Robert H. White-Stevens. Most Americans watching the show assumed that Carson, with her "poison book" in hand, would be strident, sensational, and wild-eyed, while White-Stevens, who wore a lab coat, would be the level-headed male voice of reason. This was, after all, the early 1960s; women scientists were virtually nonexistent. It didn't work out that way. "If man were to faithfully follow the teachings of Miss Carson," said White-Stevens, with a flourish of hyperbole, "we would return to the Dark Ages, and the insects and diseases and vermin would once again inherit the earth."

Carson, on the other hand, was patient and calm. "We've heard the benefits of pesticides," she said. "We have heard a great deal about their safety, but very little about their hazards, very little about their failures, their inefficiencies. And yet the public was being asked to accept these chemicals, was being asked to acquiesce in their use, and did not have the whole picture. So I set about to remedy the balance there." Carson, not White-Stevens, was given the last word. "We still talk in terms of conquest," she said. "We still haven't become mature enough to think of ourselves as only a tiny part of a vast and incredible universe. I think we're challenged as mankind has never been challenged before, to prove our maturity and our mastery, not of nature, but of ourselves." Fifteen million Americans watched *CBS Reports*. Rachel Carson had become a phenomenon. A few weeks later, Carson appeared on the *Today Show,* yet another opportunity to warn millions of Americans about the dangers of pesticides.

The day after *CBS Reports* aired, Senator Hubert Humphrey (D-MN) asked the Committee on Government Operations to conduct a congressional review of environmental hazards, including pesticides. On May 15, 1963, Rachel Carson appeared as the star witness. At issue was not whether there would be broader federal oversight of pesticide use, but rather which agency would do it. Jostling for control were the Food and Drug Administration, the U.S. Department of Agriculture, the Department of the Interior, and the Department of Health, Education, and Welfare. When Carson sat down in front of the microphone, Senator Abraham Ribicoff (D-CT), the chair of the committee, enthused, "You're the lady who started it all!" After Carson finished testifying, Senator Ernest Gruening (D-AK) predicted that *Silent Spring* would "change the course of history." Two days after the Ribicoff committee meeting, Carson appeared before the Department of Commerce and asked for a "Pesticide Commission" to oversee the use of pesticides. Ten years later, Carson's "Pesticide Commission" became the Environmental Protection Agency.

Rachel Carson had become so famous, so well known, and so sought after that a typical day included being the subject of a *Peanuts* cartoon in the morning and a call to the White House in the afternoon. (In the *Peanuts* cartoon, Lucy is talking to Schroeder, the little boy who plays the piano. "Rachel Carson says that when our moon was born, there were not oceans on Earth," says Lucy. "Rachel Carson! Rachel Carson! Rachel Carson!" shrieks Schroeder. "You're always talking about Rachel Carson!" "We girls need our heroines," Lucy replies.)

In the year and a half between the publication of *Silent Spring* and her death, Rachel Carson received the Conservation Award from the Izaak Walton League of America, the Audubon Medal from the National Audubon Society, the Cullum Geographical Medal from the American Geographical Society, the Schweitzer Medal from the Animal Welfare Institute, the Woman of Conscience Award from the Women's National Book Association, and the Conservationist of the Year Award from the National Wildlife Federation. Later, she became one of only four women elected to the American Academy of Arts and Letters, linking her to literature's immortals.

Rachel Carson's influence continued long after her death in 1964. Seventeen years later, in 1981, she won the Presidential Medal of Freedom. Two decades after that, Al Gore, who would ignite another firestorm with his movie about global warming titled *An Inconvenient Truth,* paid tribute to "the mother of the environmental movement." "*Silent Spring* came as a cry in the wilderness," said Gore, "a deeply felt, thoroughly researched, and brilliantly written argument that changed the course of history. Without this book, the environmental movement might have been long delayed or never have developed at all."

Today, most people under the age of 40 have probably never heard of Rachel Carson. But in the early 1960s, almost every American knew her name.

WHAT HARRIET BEECHER STOWE'S *Uncle Tom's Cabin* did for civil rights legislation, and Upton Sinclair's *The Jungle* did for food and drug legislation, Rachel Carson's *Silent Spring* did for environmental legislation.

On January 1, 1970—six years after Rachel Carson had died—President Richard Nixon signed the National Environmental Policy Act into law, declaring that, "the environmental decade was at hand."

In quick succession, legislators created the Council on Environmental Quality; the Environmental Protection Agency; the Occupational Safety and Health Administration; the Clear Air Act; the Clean Water Act; the Federal Insecticide, Fungicide, and Rodenticide Act; the Safe Drinking Water Act; the Environmental Pesticides Control Act; the Toxic Substances Control Act; and the Endangered Species Act.

On April 22, 1970, millions of Americans and tens of millions more around the world celebrated the first Earth Day. Conservationism had become environmentalism. Groups like the Sierra Club (founded in 1892), the National Audubon Society (founded in 1905), the World Wildlife Fund (founded in 1947), and the Nature Conservancy (founded in 1951) were conservationists, intent on protecting natural resources and improving national parks. These new environmental groups, like Clean Water Action and the Natural Resources Defense Council, were different: more passionate, more confrontational, and less forgiving. Now the focus was on protesting pollution and cleaning the air and water. Now there were the good guys and the bad guys. And the bad guys, like the chemical industry, weren't going to be allowed to get away with it any longer. Born of Rachel Carson's book, these new groups scared the older ones. None of the conservation societies participated in the first Earth Day.

When the dust settled, Rachel Carson was a hero, the unquestioned goddess of a movement that has only gained momentum in the 50 years since publication of her game-changing book. "The Rachel Carson we think of as the author of *Silent Spring,*" wrote one acolyte, "[was the] birth mother of modern environmentalism: messenger of a story that rocked the world. The real Rachel Carson never met her . . . She didn't live long enough to become acquainted with the Carson we know, that towering figure whose light illuminated our sense of the world forever."

ALTHOUGH RACHEL CARSON'S *Silent Spring* shined a long-overdue light on our indiscriminate use of pesticides, it had its flaws. Not everyone loved *Silent Spring*.

Some of the criticism came from writers. *Time* magazine decried Carson's penchant for overstatement: "Scientists, physicians and other technically informed people will also be shocked by *Silent Spring*—but for a different reason. They recognize Miss Carson's skill in building her frightening case; but they consider that case unfair, one-sided, and hysterically overemphatic. Many of the scary generalizations—and there are lots of them—are patently unsound."

Other criticisms came, not unexpectedly, from the chemical industry. Velsicol, at the time one of the world's leading manufacturers of pesticides such as chlordane, heptachlor, and endrin, threatened to sue Houghton Mifflin, the publisher of *Silent Spring,* for libel. A few months later, however, Velsicol's attention was diverted when five million fish turned belly up in the lower Mississippi River, a consequence of a massive endrin contamination from one of its treatment plants.

None of these criticisms or threats was surprising. One criticism, however, was. It came from Luther Terry, the surgeon general of the United States. Terry was worried that by making DDT synonymous with *poison,* the world was about to lose a powerful weapon in the fight against some of its biggest killers. He had reason for concern.

DDT HAS A LONG and rich history.

In 1874, Othmar Zeidler, a graduate student at the University of Strasbourg in Germany, was looking to create a new substance for his thesis. He combined chloral hydrate with chlorobenzene in the presence of sulfuric acid. The result was DDT (dichlorodiphenyltrichloroethane). Zeidler didn't study DDT's properties. He didn't care about its properties. He just wanted to create a new substance so he could graduate. As a consequence, DDT sat on the shelf for 65 years.

In 1939, Paul Müller, an employee of the J. R. Geigy Company in Basel, Switzerland, was working on a method to kill clothes moths without damaging clothes. Müller stumbled upon Zeidler's formula. What he found surprised him. Not only did DDT kill the moths, it also killed flies, mosquitoes, lice, and ticks—insects responsible for transmitting some of the world's deadliest diseases. Better still, DDT's killing power seemed to last for months.

AT THE START OF WWII, knowing that wars spread disease, J. R. Geigy released its formula for DDT to the Germans and the Allies. The Germans ignored it; the Americans and Brits didn't. In America, the Cincinnati Chemical Works was the first to mass-produce it. Soon, 14 other American companies and several British companies joined in. Production couldn't have come at a better time. The reason: typhus.

Typhus is a bacterium spread by the body louse. The bacterium *(Rickettsia prowazekii)* is named for the two researchers who discovered it—Howard Ricketts and Stanislaus von Prowazek, both of whom died from the disease. The lice deposit their feces, which contain the typhus bacteria, onto the skin. The intense itching that invariably follows allows bacteria to penetrate the skin and enter the bloodstream, causing chills, fever, headache, rash, coma, and death. During WWII, more people died from typhus than from combat.

In January 1944, DDT made its debut in Naples, Italy, a city in the midst of a massive typhus epidemic. After setting up delousing stations, the Allies sprayed DDT onto 72,000 Italian citizens every day—more than 1.3 million people in all. Within three weeks, the outbreak was under control. By the end of 1944, factories were producing more than a million pounds of DDT every month. With their new weapon in hand, health officials dusted millions of soldiers, fogged military barracks, and sprayed whole islands to protect the Marines before they landed. By 1945, DDT production had reached 36 million pounds a year. Because neither the Germans nor the Japanese used it, some have argued that DDT helped the Allies win the war.

DDT was also used to delouse concentration camp survivors. One dramatic story involved the prison camp at Bergen-Belsen. At the time of liberation, in 1945, typhus had infected more than 20,000 camp prisoners. When the British soldiers who liberated the camp first started spraying survivors, most of the prisoners were skeptical. "After two to three days at the hospital," recalled one, "we have our first encounter with the pesticide DDT. When the English soldiers enter the hospital room with sprayers filled with this product, we all look at them with contemptuous superiority. They're planning on using this puny white powder to destroy all these millions of lice?! Yet, right in front of our eyes, something close to a miracle starts to happen. Slowly, the incessant itching, so painful on our pus-infected, ulcerated skin, starts to vanish, and this great relief finally convinces us that we really have been liberated." (Liberation came too late for one Bergen-Belsen prisoner, Anne Frank, who died from typhus.)

In 1948, for his work demonstrating DDT's benefits to public health, Paul Müller won the Nobel Prize in medicine and physiology.

ALTHOUGH TYPHUS was a killer, it paled in comparison to the infection that has killed—and continues to kill—more people than any other: malaria. Spread by the bite of the anopheles mosquito, the malaria parasite infects the liver and blood, causing high fever, shaking chills, bleeding, disorientation, and death. In 1962, when Rachel Carson published *Silent Spring,* the best weapon to control malaria wasn't drugs like quinine and chloroquine, or environmental measures like mosquito nets or swamp drainage. Arguably, the best, cheapest, and most effective weapon in the fight against malaria was DDT. Following a spraying program in South Africa, the number of malaria cases decreased from 1,177 cases in 1945 to 61 cases in 1951; in Taiwan, from more than a million cases in the mid-1940s to 9 cases in 1969; and, in Sardinia, from 75,000 cases in 1946 to 5 cases in 1951.

Malaria also hit close to home. In the early 1900s, more than a million Americans were infected with malaria every year. Although improvements in housing, better standards of living, and control of mosquito breeding sites had clearly lessened the incidence of the disease, DDT spraying was enormously beneficial, especially in rural areas. Between January 1945 and September 1947—as part of a program run by the MCWA (Malaria Control in War Areas)—more than three million houses were sprayed in the Southeast. In 1952, the United States was finally declared free of malaria. (Located in Atlanta, Georgia, the MCWA later changed its name to the Centers for Disease Control.)

In 1955, the World Health Assembly directed the World Health Organization to launch a global malaria elimination program with DDT as its centerpiece. By 1959, when the program swung into operation, more than 300 million people had already been saved by DDT. By 1960, malaria had been eliminated from 11 countries. As malaria rates went down, life expectancies went up, as did crop production, land values, and relative wealth. Probably no country benefited more from the WHO program than Nepal, where spraying began in 1960. At the time, more than two million Nepalese, mostly children, suffered from malaria. By 1968, the number was reduced to 2,500. Before the malaria control program, life expectancy in Nepal was 28 years; by 1970, it was 42 years.

MALARIA WASN'T THE ONLY disease transmitted by the bite of a mosquito. DDT also dramatically reduced the incidence of yellow fever and dengue. Furthermore, DDT killed fleas, like the ones that lived on rats that transmitted murine typhus, and the ones that lived on prairie dogs and ground squirrels and transmitted *Yersinia pestis,* the plague. Considering the virtual elimination of all of these diseases in many countries, the National Academy of Sciences estimated in 1970 that DDT had saved the lives of 500 million people. One could argue reasonably that DDT has saved more lives than any other chemical in history.

ENVIRONMENTALISTS DIDN'T SEE IT that way. Inspired by Rachel Carson's *Silent Spring,* they targeted DDT for elimination. In 1969, Wisconsin and Arizona banned DDT; so did Michigan, which published a formal obituary in a local newspaper: "Died. DDT, age 95: a persistent pesticide and onetime humanitarian. Considered to be one of World War II's greatest heroes, DDT saw its reputation fade after it was charged with murder by author Rachel Carson. Death came on June 27 in Michigan after a lingering illness. Survived by dieldrin, aldrin, endrin, chlordane, heptachlor, lindane, and toxaphene. Please omit flowers." Ironically, every one of these surviving chemicals was far more dangerous to human health than DDT.

Sensing the public's fear of pesticides, President Richard Nixon promised to ban DDT from the United States by the end of 1970, even though the Department of Agriculture didn't believe an adequate substitute was available. In 1972, William Ruckelshaus, head of the newly created Environmental Protection Agency—against strong opposition from the Pan American Health Organization, the World Health Organization, and many public health advocates in the United States—banned DDT from use in the United States. Other countries followed. Public health officials, sensing the disaster that was about to unfold, urged countries that made DDT to continue to make it. But it was too late. By the mid-1970s, under pressure from environmental groups, support for international DDT programs had dried up.

Those inspired by *Silent Spring* had spared mosquitoes from the killing effects of DDT. But they hadn't spared children from the killing effects of mosquitoes.

USING DDT AS A LADDER, the United States had climbed out of the cesspool—ridding itself of anopheles mosquitoes; no longer would its citizens have to suffer malaria. Then, in the name of environmentalism, Americans pulled the ladder up behind them, leaving developing world countries the options of using biological strategies that didn't work or antimalarial drugs they couldn't afford.

Since 1972, when the Environmental Protection Agency banned DDT from the United States, about 50 million people have died from malaria: Most have been children less than five years old.

Examples of the impact of *Silent Spring* abound:

In India, between 1952 and 1962, DDT spraying caused a decrease in annual malaria cases from 100 million to 60,000. By the late 1970s, no longer able to use the pesticide, the number of cases increased to 6 million.

In Sri Lanka, before the use of DDT, 2.8 million people suffered from malaria. When the spraying stopped in 1964, only 17 people suffered from the disease. Then, between 1968 and 1970, no longer able to use DDT, Sri Lanka suffered a massive malaria epidemic—1.5 million people were infected by the parasite.

In South Africa, where DDT use was banned in 1997, the number of malaria cases increased from 8,500 to 42,000 and malaria deaths from 22 people to 320.

In the end, 99 countries eliminated malaria; most used DDT to do it. "Banning DDT is one of the most disgraceful episodes in twentieth century America," wrote author Michael Crichton. "We knew better and we did it anyway and we let people around the world die, and we didn't give a damn."

ENVIRONMENTALISTS HAVE ARGUED that when it came to DDT, it was pick your poison. If DDT was banned, more people would die from malaria. But if DDT wasn't banned, then people would suffer and die from a variety of other diseases, not the least of which were leukemia and other cancers. There was one problem with this line of reasoning: Despite Carson's warnings in *Silent Spring,* studies in Europe, Canada, and the United States showed that DDT didn't cause liver disease, premature births, congenital defects, leukemia, or any of the other diseases she had claimed. Indeed, the only type of cancer that had increased in the United States during the DDT era was lung cancer, which was caused by cigarette smoking. DDT was arguably the safest insect repellent ever invented—far safer than many of the other pesticides that have since taken its place.

Still, environmentalists argued that we aren't alone on this planet. We share it with many other species. Aren't we responsible for them, too? The final irony of *Silent Spring* was that Rachel Carson hadn't only overstated DDT's effects on human health; she had overstated its effects on animal health.

RACHEL CARSON ORIGINALLY called her book, *Man Against Nature.* But her agent, Marie Rodell, didn't think that was poetic enough. So she presented Carson with a line from the English Romantic poet John Keats's "La Belle Dame Sans Merci" ("The Beautiful Lady Without Mercy"): "The sedge is withered from the lake / and *no birds sing.*"

*Silent Spring* was born. Carson's prose was unequivocal; DDT was killing birds.

But the evidence wasn't clearly on her side.

Every winter, the National Audubon Society performs its Christmas bird counts. Between 1941, before DDT, and 1960, after DDT had been used for at least a decade, 26 different kinds of birds had been counted. All had increased in number. In *Silent Spring,* Carson focused on specific instances where DDT had damaged starlings, robins, meadowlarks, and cardinals. But, at least according to the Christmas counts, populations of each of these birds had actually increased about fivefold.

Another bird targeted by DDT—a symbol of America's strength and freedom—was the eagle. "Like the robin," wrote Carson, "another American bird appears to be on the verge of extinction. This is the national symbol: the eagle. Its population has dwindled alarmingly with the past decade." As proof, Carson cited the findings of Charles Broley, a retired banker who lived on Florida's west coast who had noticed that the number of bald eagle nests between Tampa and Fort Myers had declined. What Carson had failed to mention was that this decline had occurred *before* DDT was used (prior to 1940), and was due to habitat destruction and killing by hunters, either for sport or to protect livestock. In fact, between 1939 and 1961, during the time of heaviest DDT use, the Christmas counts had shown an *increase* in eagle populations. The reason: the Eagle Protection Act of 1940, which prohibited the hunting, capturing, and killing of the birds. During the ten years before DDT was banned, the number of bald eagle nesting pairs had doubled.

That bird populations were actually increasing during the period of heaviest use of DDT wasn't a coincidence. DDT was beneficial in that it protected birds from a broad range of insect-borne diseases such as malaria, Newcastle disease, encephalitis, rickettsialpox, and bronchitis, and—because DDT lessened the harmful effects of pests on crops—it made more seeds and fruits available for birds to eat.

Rachel Carson wasn't only a member of the National Audubon Society; she had also participated in its annual Christmas bird counts. So she must have known about the bird population data; still, she had chosen to ignore them. In *Silent Spring,* Rachel Carson never mentioned habitat destruction, egg collection, or hunting as reasons for why bird populations might have dwindled. It was a pesticide witch hunt. "Readers of *Silent Spring,* in the 1960s and even now, are impressed by its poetic language and imagery, but it did not escape the notice of scientists that while the book was heavy on prose it was light on science," wrote Donald Roberts and colleagues in *The Excellent Powder: DDT's Political and Scientific History*. "It seems certain that scientists and students of chemistry and the natural world could not have guessed how *Silent Spring* would pave the way for science to be sidelined in the development of laws, policies, and global strategies for disease control."

When the EPA banned DDT in the early 1970s, much of the information about whether it had caused human disease or affected wildlife was already available. This information came to light at a public hearing forced by the Environmental Defense Fund (EDF). EDF officials wanted the public, the press, and politicians to hear just how harmful DDT could be. So they called a variety of environmentalists to testify on their behalf. Health officials, however, didn't take this attempt at a public flogging lying down. They called their own experts in the fields of chemistry, toxicology, agriculture, and environmental health.

The hearing lasted eight months, included 125 witnesses and 365 exhibits, and generated a transcript that was 9,312 pages long. When it was over, Edward Sweeney, the hearing examiner, rendered his verdict: "DDT is not mutagenic [causing cancer] or teratogenic [causing birth defects] to man," he wrote. "The uses of DDT under the registration involved here do not have a deleterious effect on freshwater fish, estuarine organisms, wild birds or other wildlife. The [EDF has] not fully met the burden of proof. *There is a present need for the continued use of DDT for the essential uses defined in this case.*" William Ruckelshaus, head of the newly formed Environmental Protection Agency, never attended the meeting. When it was over, he never read the report. Rather, on June 2, 1972, Ruckelshaus unilaterally banned DDT. It was a political decision, yielding to public sentiment. And it ignited an international firestorm against DDT that resulted in it being banned from the world.

The chemical industry seemed not to care. DDT was just one of many pesticides used in agriculture. And the agricultural market was far more lucrative than the public health market. Now, DDT could be replaced with drugs that were not only more expensive, but far more harmful to people.

IN MANY WAYS, Rachel Carson had sounded an important alarm. She was the first to say that we needed to be more attentive to our impact on the environment. (Indeed, climate change has been a direct consequence of man-made activities.) She was the first to warn us that DDT could accumulate in the environment. (Even after the spraying stopped, DDT and its by-products were still present throughout the ecosystem.) And she was right in her prediction that biological insect controls might eventually be of value. (Decades after publication of *Silent Spring*, the bacterium *Bacillus thuringiensis israelensis* [Bti], which kills mosquito larvae, was included in malaria eradication efforts.) Unfortunately, Rachel Carson had taken one step too far. By claiming that DDT caused diseases like leukemia in children—or by claiming that children could be fine one minute and dead a few hours later—she had scared the hell out of the American public. In the end, Rachel Carson wasn't the scientist she had claimed to be. She was a polemicist, willing to stretch the truth to fit her bias.

*SILENT SPRING* WAS SUCCESSFUL because it was lyrical, compelling, and dramatic. But there was another reason it had had such an enormous impact: *Silent Spring* was biblical, appealing to our notion that we had sinned against our creator.

The book begins in Eden. "There was once a town in the heart of America where all life seemed to live in harmony with its surroundings." Man, however, had eaten from the tree of knowledge, worshipping the false god of economic progress while destroying paradise. As a consequence, "a shadow of death had fallen on the people and the land." And so man was to be cast out of Eden, forced to toil on a scorched earth while suffering all manner of illnesses.

In truth, Rachel Carson's Eden never existed. And nature has never been in balance. It's been in constant flux, arguably in chaos. Because the simple truth is that Mother Nature isn't much of a mother: She can kill us, and unless we fight back, she will. "[Carson] paints a nostalgic picture of Elysian life in an imaginary American village of former years, where all was in harmonious balance with Nature and happiness and contentment reigned interminably," wrote one scientist. "But the picture she paints is illusory. [T]he rural Utopia she describes was rudely punctuated by a longevity among its residents of perhaps thirty-five years, by an infant mortality of upwards of twenty children dead by the age of five of every hundred born, by mothers dead in their twenties from childbed fever and tuberculosis, by frequent famines crushing isolated peoples through long, dark, frozen winters following the failure of a basic crop the previous summer, [and] by vermin and filth infesting their homes . . . Surely she cannot be so naïve as to contemplate turning our clocks back to the years when man was indeed immersed in Nature's balance and barely holding his own."

William Cronon, an environmental scientist and the author of *Changes in the Land,* took Carson's argument to its illogical end: "It is not hard to reach the conclusion that the only way human beings can hope to live naturally on the earth is to follow the hunter-gatherers back into a wilderness Eden and abandon virtually everything that civilization has given us. If nature dies because we enter it, then the only way to save nature is to kill ourselves." Biologist I. L. Baldwin sounded a similar theme: "Modern agriculture and modern public health, indeed, modern civilization could not exist without a relentless war against the return of a true balance of nature." Carson never saw it that way, insisting on a world that had never existed: "Under primitive agricultural conditions the farmer had few insect problems," she wrote, ignoring that fact that early farming societies were riddled with insect-borne diseases and insect-induced famines.

In 2006, the World Health Organization, realizing its mistake, changed its position on DDT, no longer bowing to political pressures to ban the product. On September 15, Dr. Arata Kochi, director of the Global Malaria Programme, announced the new policy: "I asked my staff. I asked malaria experts around the world, 'Are we using every possible weapon to fight this disease?' It became apparent that we were not. One powerful weapon against malaria was not being deployed. In a battle to save the lives of nearly one million children a year—most of them in Africa—the world was reluctant to spray the inside of houses and huts with insecticides; especially with the highly effective insecticide known as dichlorodiphenyltrichloroethane or DDT." The Sierra Club backed Kochi; the Pesticide Action Network didn't.

For more than 30 years, countries where malaria epidemics were common had been denied this lifesaving chemical. Although there were alternatives, and some of those alternatives were used, no chemical was as cheap, long lasting, or effective as DDT. As a result, millions of people, mostly children, died needlessly.

Carson's supporters have heard the criticisms. They've argued that, had she lived longer, she would never have promoted a ban on DDT. Indeed, in *Silent Spring,* Carson wrote, "It is not my contention that chemical pesticides never be used." But it *was* her contention that DDT had caused leukemia, liver disease, birth defects, premature births, and a whole range of chronic illnesses. An influential author cannot, on the one hand, claim that DDT causes leukemia (which, in 1962, was a death sentence) and then, on the other hand, expect that anything less than a total ban on the chemical would be the result.

"The question is whether any civilization can wage relentless war on life without destroying itself, and without losing the right to be called civilized," wrote Rachel Carson in *Silent Spring*. Roger Meiners, co-author of *Silent Spring at 50: The False Crises of Rachel Carson,* countered, "This rhetorical question suggests another: whether any civilization that hobbles new technology that could reduce hunger and disease, on the chance that the new technology might have negative consequences—essentially giving up a real bird in hand for a hypothetical bird in the bush—should lose the right to be called civilized."

The lesson from Rachel Carson and the banning of DDT reprises an earlier theme—**it's all about the data**—as well as suggesting two new ones.

When officials at the Environmental Protection Agency (EPA) were deciding whether to ban DDT, they had two sets of data from which to choose. One was a 9,000-page report generated by more than a hundred experts in the fields of chemistry, toxicology, agriculture, and environmental health that included hundreds of graphs and figures. DDT, the report concluded, wasn't killing birds, wasn't killing fish, and wasn't causing chronic diseases in people. Although numbingly boring, the report was accurate.

The other source of evidence was a book: Rachel Carson's *Silent Spring*—a beautifully written, heart-pounding tale with biblical overtones. Unlike the expert report, however, it was short on data and long on anecdotes. For example, to prove that eagles were dying from DDT, Carson had relied on the observations of a retired banker from Florida whose hobby was bird-watching. In the end, the EPA's decision to ban DDT wasn't based on data; it was based on fear and misinformation.

Carson's story provides another lesson. In the 16th century Paracelsus, a Swiss physician and philosopher said, "**the dose makes the poison**." When Rachel Carson wrote *Silent Spring,* she appealed to a 1960s, back-to-nature mentality supported by young, energetic, community-minded activists. Carson's basic premise—that man-made activities were destroying the environment—was correct. Thanks to Rachel Carson, we are now far more attentive to our impact on the planet. Unfortunately, Carson also gave birth to the notion of zero tolerance—the assumption that any substance found harmful at any concentration or dosage should be banned absolutely. If large quantities of DDT (like those used in agriculture) were potentially harmful, then even small quantities (like those used to prevent mosquitoes from biting) should be avoided. In a sense, Rachel Carson was an early proponent of the precautionary principle. But, as we'll see in the final chapter with cancer-screening programs, we should **be cautious about being cautious**.

# Chapter 7

# NOBEL PRIZE DISEASE

*"Pride goes before destruction, haughty spirit before a fall."*

—Proverbs 16:18

Vitamin manufacturers today owe their multibillion-dollar-a-year business to one man: a Nobel Prize–winning scientist who, when he wandered far outside of his field, caused us to believe that large quantities of supplemental vitamins would make us live longer, better, healthier lives. In fact, they have only increased our risks of cancer and heart disease.

Linus Pauling was a genius.

In 1931, Pauling published a paper in the *Journal of the American Chemical Society* titled, "The Nature of the Chemical Bond." At the time, chemists had already described two different types of bonds: ionic (where one atom gives up an electron to another) and covalent (where atoms share electrons). Pauling said that it didn't have to be one or the other—there was something in between. It was a novel and shocking concept—for the first time marrying quantum physics with chemistry. Pauling's description of chemical bonding was so revolutionary, so far ahead of its time, that the editor of the journal had trouble finding an expert qualified to review it. "It was too complicated for me," said Albert Einstein.

For this single paper, Linus Pauling was awarded the Langmuir Prize as the most outstanding chemist in the United States, elected to the National Academy of Sciences—the single highest honor that can be bestowed on a scientist by his peers—and made a full professor at Caltech, one of the most prestigious universities for science and engineering in the world. He was only 30 years old. And he was just getting started.

In 1949, Pauling published a paper in *Science* titled, "Sickle Cell Anemia: A Molecular Disease." At the time, scientists knew that people with sickle-cell disease suffered crippling pain when their red blood cells changed from plump round disks to thin narrow sickles. What they didn't know was why. Pauling showed that hemoglobin, the molecule in red blood cells that carries oxygen from the lungs to the rest of the body, had a slightly different electrical charge in patients with sickle-cell disease. It was the first time that a scientist had described the molecular basis of a disease, launching the field of molecular biology.

In 1951, Pauling published a paper in the *Proceedings of the National Academy of Sciences* titled "The Structure of Proteins." Taking yet another Einsteinian leap, Pauling showed that proteins folded upon themselves in recognizable patterns. At the time of publication, scientists knew that proteins were made of a series of linked amino acids. But they hadn't envisioned what proteins looked like in three dimensions. Pauling did. One of the protein structures Pauling described was called the alpha helix, a finding that allowed James Watson and Francis Crick to solve the structure of DNA: nature's blueprint.

In 1954, for his work on chemical bonding and protein structure, Linus Pauling won the Nobel Prize in chemistry.

Pauling was also active outside the laboratory. Throughout the 1950s and 1960s, Linus Pauling became one of the world's most recognizable peace activists. He opposed the creation of the atomic bomb and forced government officials to admit that nuclear radiation damaged human DNA. His efforts were rewarded with the first nuclear test ban treaty. They were also rewarded with his second Nobel Prize, this time for peace. Linus Pauling had become the first (and so far only) person in history to win two unshared Nobel Prizes. In 1961, Pauling appeared on the cover of *Time* magazine, hailed as one of the greatest scientists who had ever lived.

Then, in the mid-1960s, Linus Pauling fell off an intellectual cliff.

To THOSE WHO KNEW HIM, Pauling's lack of rigor wasn't surprising. It had first appeared in his science.

In 1953, Pauling published a paper in the *Proceedings of the National Academy of Sciences* titled "A Proposed Structure for the Nucleic Acids." Pauling claimed that DNA was a triple helix. (Within a year, Watson and Crick proposed their now famous double-helix model.) It was the single greatest scientific error of his career. And his colleagues never let him forget it. Whereas Pauling had spent decades considering the structure of proteins, he had spent only a few months on the structure of DNA. His wife, Ava Helen, later remarked, "If that was such an important problem, why didn't you work harder on it?" James Watson was less kind, remembering his surprise "that a giant had forgotten elementary college chemistry." "If a student had made a similar mistake," said Watson, "he would be thought unfit to benefit from Caltech's chemistry department," where Pauling was a professor.

But Linus Pauling's full descent into the abyss began on a single day in March 1966, when he was 65 years old. Pauling was in New York City where he had just accepted the Carl Neuberg Medal for his scientific achievements. During his talk, Pauling said that he wished only that he could live another 25 years so he could see how certain scientific investigations were proceeding. Pauling later wrote, "On my return to California, I received a letter from a biochemist, Irwin Stone, who had been at the talk. He wrote that if I followed his recommendation of taking 3,000 milligrams of vitamin C, I would live not only 25 years longer, but probably more."

Pauling followed Stone's advice, taking 10, then 20, then 300 times the recommended daily allowance of vitamin C, eventually 18,000 milligrams a day. It worked. Pauling said that he felt livelier, healthier, and better than ever before. No longer did he have to suffer the debilitating colds that had plagued him for years. Convinced that he had stumbled upon the fountain of youth, Linus Pauling, with the weight of two Nobel Prizes behind him, became the nation's leading advocate for megavitamins. Based on his limited personal experience, Pauling recommended megavitamins and various dietary supplements for mental illness, hepatitis, polio, tuberculosis, meningitis, warts, strokes, ulcers, typhoid fever, dysentery, leprosy, fractures, altitude sickness, radiation poisoning, snakebites, stress, rabies, and virtually every other disease known to man. Now a zealot for a cause, Linus Pauling would later ignore study after study showing that he was wrong. Clearly and spectacularly wrong.

THE MEETING BETWEEN Linus Pauling and Irwin Stone was a watershed moment in the history of the vitamin and supplement craze in the United States—made all the more remarkable by the contrast between the two men. Pauling was the product of a classical education, well grounded in the fields of chemistry and physics. Stone, who was generously described by Pauling as a "biochemist," had studied chemistry for two years in college before receiving an honorary degree from the Los Angeles College of Chiropractic and a bogus Ph.D. from Donsbach University, a nonaccredited correspondence school in California. Pauling had succeeded in unlocking some of nature's best kept secrets because he was dogged in his devotion to formal proofs—the kind of proofs that result in publications in major scientific journals and the kind of proofs that win Nobel Prizes. Stone had never received a valid scientific credential, never published a paper in a medical or scientific journal, and had graduated from a program in Los Angeles that taught that all human diseases were the result of misaligned spines. Yet Pauling accepted Stone's revelations uncritically.

In 1970, Linus Pauling published his first book, *Vitamin C and the Common Cold,* which urged Americans to take 3,000 milligrams of vitamin C every day—roughly 500 times the recommended daily allowance. The book became a national best seller. Within a few years, more than 50 million Americans—1 of every 4 people living in the United States—were following Pauling's advice. Scientific studies, however, failed to support him.

In 1942, about 30 years *before* Pauling published his book on vitamin C, a group of researchers from the University of Minnesota published a study in the *Journal of the American Medical Association* of 980 people with colds, finding that vitamin C did nothing to lessen symptoms.

After Pauling published his book, and largely in response to its popularity, researchers at the University of Maryland and the University of Toronto and in the Netherlands performed several studies of volunteers who had been given 2,000, 3,000, or 3,500 milligrams of vitamin C a day for the prevention or treatment of colds. Again, large doses of vitamin C were found to be useless.

Because of these and other studies, not a single professional medical, scientific, or public health organization recommends vitamin C for the prevention or treatment of colds. Unfortunately, it's been hard to unring the bell. Once Pandora's box is opened, you cannot put anything back inside; once Americans had become convinced that vitamin C was a wonder drug, there was no going back.

Then Linus Pauling doubled down, claiming that vitamin C also cured cancer.

IN 1971, PAULING WROTE that megadoses of vitamin C (those greatly in excess of the recommended daily allowance) would cause a 10 percent decrease in the incidence of cancer in the United States; six years later, he upped his prediction to 75 percent. If we followed his advice, Pauling believed that vitamin C could make us practically immortal, living longer than ever before. He predicted that the average American life span would increase to a hundred years, then 150 years. Like *Vitamin C and the Common Cold,* his books *Cancer and Vitamin C* and *How to Live Longer and Feel Better* also became instant best sellers. Linus Pauling was so powerful, such a media darling, that cancer victims started to take his advice. Doctors, blindsided by Pauling's influence, had no choice but to see if he was right.

In 1979, Charles Moertel and colleagues at the famed Mayo Clinic in Rochester, Minnesota, studied 150 cancer victims. Half were given 10,000 milligrams of vitamin C a day (roughly 1,500 times the recommended daily allowance) and half weren't. They published their paper, titled "Failure of High-Dose Vitamin C Therapy to Benefit Patients with Advanced Cancer: A Controlled Trial," in the *New*

*England Journal of Medicine*. The title said it all; vitamin C hadn't worked. Pauling was incensed. Surely Moertel hadn't done the study correctly. Then Pauling found what he believed was the flaw in the experiment: Moertel had given vitamin C to patients who had already received chemotherapy, negating its wondrous healing properties. Pauling was now convinced that vitamin C worked only in patients who hadn't received any chemotherapy.

Although he didn't really see the point, Moertel was bullied into performing another study of vitamin C in cancer victims, this time in patients who had yet to receive chemotherapy. In 1985, he published his second study, again in the *New England Journal of Medicine* and again showing no difference. Now Pauling was really angry, accusing Moertel of "deliberate fraud and misrepresentation." He considered suing Moertel, but his lawyers talked him out of it.

Linus Pauling had been so right for so long that he just couldn't imagine that he could ever be wrong—even when he clearly *was* wrong. As described by biographers and colleagues, Pauling's failures could have been predicted from his personality. "Linus Pauling is a classic example of a person who loves humanity but doesn't care much for people," wrote biographers Ted and Ben Goertzel. "He is generally without close friends. Politically, he is a crusader for his vision of truth with little tolerance for considering the viewpoints of others." Like the Goertzels, Max Perutz—a colleague of Pauling's who had also won a Nobel Prize in chemistry—praised Pauling for his breakthrough work, but also alluded to a darker side: "It seems tragic that [vitamin C] should have become one of Pauling's major preoccupations during the last twenty-five years of his life and spoilt his great reputation as a chemist. Perhaps it was related to his greatest failing: his vanity. When anybody contradicted Einstein, he thought it over, and if he found he was wrong, he was delighted, because he felt he had escaped from an error. But Pauling would *never* admit that he might have been wrong. When, after reading Pauling and [Robert] Corey's paper on the alpha helix, I discovered [a problem with their calculations], I thought he would be pleased. But no, he attacked me furiously, because he could not bear the idea that someone else had thought of a test for the alpha helix of which he had not thought himself."

AMONG PAULING'S MISANTHROPIC dealings with those who dared to oppose him—dared to believe that he could ever be wrong—no story was sadder or more telling than that of Arthur Robinson.

In 1973, Pauling founded the Institute of Orthomolecular Medicine in Menlo Park, California, later to become the Linus Pauling Institute. His biggest supporter was the pharmaceutical giant Hoffman-La Roche, one of the world's largest manufacturers of vitamins and dietary supplements. Pauling decided that if other researchers were unable to show that megavitamins were wonder drugs, then he would do it himself.

When Pauling founded his institute, he brought Arthur Robinson along with him. Pauling was president, director, and chairman of the board. Robinson, a chemist and one of the brightest students to have ever graduated from the University of California in San Diego, was vice president, assistant director, and treasurer. Robinson's job was to provide experimental evidence for Pauling's theories about vitamin C. It didn't work out that way.

In 1977, Arthur Robinson evaluated a special breed of mice that suffered from skin cancer. To some he gave the human equivalent of 10,000 milligrams of vitamin C a day; to others, he didn't give any extra vitamins. The results were alarming. Robinson found that high doses of vitamin C actually *increased* their risk of cancer.

Robinson knew that Pauling and his wife were taking large doses of vitamin C. Concerned, he told Pauling of his results. "At that time [1970]," recalled Robinson, "he had put himself and his wife on at least 10,000 milligrams a day of vitamin C, and they were on it for the next decade. I pointed out that she was bathing her stomach with an enormous amount of mutagenic [cancer-causing] material for ten years." (Ava Pauling would later suffer from stomach cancer.)

Pauling refused to believe it, threatening to have the mice killed and demanding Robinson's resignation. "He claimed that his famous name gave him the right to absolute control over all ideas and research at the institute," recalled Robinson. "Linus informed me that he would have me fired disgracefully from all of my positions, including that of tenured research professor, and that he would take several other actions ruinous to my professional career if I did not agree to his demands."

Following Pauling's orders, the board of trustees withheld Robinson's salary, suspended him from the institute, and locked his files. Robinson didn't go quietly, suing Pauling and the institute for $25 million. The lawsuit dragged on for five years, costing the institute $1 million in legal fees. The case was eventually settled for $500,000.

Arthur Robinson's findings weren't limited to mice. Soon other researchers found that megavitamins increased the risk of cancer in people, too.

In 1994, the National Cancer Institute, in collaboration with Finland's National Public Health Institute, studied 29,000 Finnish men: All were smokers and all were at risk of lung cancer. The men were given large doses of vitamin E, beta-carotene (a vitamin A precursor), both, or neither. The results were the opposite of what had been expected. Those given megavitamins were actually *more* likely to die from lung cancer, not less.

In 1996, investigators from the Fred Hutchinson Cancer Research Center in Seattle studied 18,000 people who, because they had been exposed to asbestos, were—like those who smoked cigarettes—also at greater risk of lung cancer. Participants were given large doses of vitamin A, beta-carotene, both, or neither. The study ended abruptly when the safety monitors realized that those taking megavitamins had a dramatically higher rate of lung cancer (28 percent greater than those not receiving vitamins) as well as heart disease (17 percent greater).

In 2004, researchers from the University of Copenhagen reviewed 14 randomized trials involving 170,000 people given large doses of vitamins A, C, E, and beta-carotene to see whether they had a lesser incidence of intestinal cancers. As had been true for lung cancer, vitamin recipients were more likely to have intestinal cancer than those who didn't take supplemental vitamins.

In 2005, researchers from Johns Hopkins School of Medicine evaluated 19 studies involving more than 136,000 people who had taken megavitamins and found an increased risk of early death in vitamin recipients. In the same year, a study published in the *Journal of the American Medical Association* evaluated more than 9,000 people who took high doses of vitamin E to prevent cancer. Again, vitamin recipients were more likely to develop cancer and heart disease.

In 2008, a review of all existing studies of more than 230,000 people who had taken megavitamins found an increased risk of cancer and heart disease.

In 2011, researchers from the Cleveland Clinic published a study of 36,000 men who took vitamin E, selenium (a mineral), both, or neither. Those who took megadoses of vitamin E had a 17 percent greater risk of prostate cancer.

Linus Pauling was wrong about megavitamins because he had made two fundamental errors. First, he had assumed that you cannot have too much of a good thing.

Vitamins are critical to life. If people don't get enough vitamins, they suffer various deficiency states, like scurvy (not enough vitamin C) or rickets (not enough vitamin D). The reason that vitamins are so important is that they help convert food into energy. But there's a catch. To convert food into energy, the body uses a process called oxidation. One outcome of oxidation is the generation of something called free radicals, which can be quite destructive. In search of electrons, free radicals damage cell membranes, DNA, and arteries, including the arteries that supply blood to the heart. As a consequence, free radicals cause cancer, aging, and heart disease. Indeed, free radicals are probably the single greatest reason that we aren't immortal.

To counter the effects of free radicals, the body makes antioxidants. Vitamins—like vitamins A, C, E, and beta-carotene—as well as minerals like selenium and substances like omega-3 fatty acids all have antioxidant activity. For this reason, people who eat diets rich in fruits and vegetables, which are rich in antioxidants, tend to have less cancer, less heart disease, and live longer. Pauling's logic to this point is clear; if antioxidants in food prevent cancer and heart disease, then eating large quantities of manufactured antioxidants should do the same thing. But Linus Pauling had ignored one important fact: Oxidation is also required to kill new cancer cells and clear clogged arteries. By asking people to ingest large quantities of vitamins and supplements, Pauling had shifted the oxidation-antioxidation balance too far in favor of antioxidation, therefore inadvertently increasing the risk of cancer and heart disease. As it turns out, Mae West aside, you actually *can* have too much of a good thing. ("Too much of a good thing can be wonderful," said West, who was talking about sex, not vitamins.)

Second, Pauling had assumed that vitamins and supplements ingested in food were the same as those purified or synthesized in a laboratory. This, too, was incorrect. Vitamins are phytochemicals, which means that they are contained in plants (*phyto-* means "plant" in Greek). The 13 vitamins (A, $B_1$, $B_2$, $B_3$, $B_5$, $B_6$, $B_7$, $B_9$, $B_{12}$, C, D, E, and K) contained in food are surrounded by thousands of other phytochemicals that have long and complicated names like flavonoids, flavonols, flavanones, isoflavones, anthocyanins, anthocyanidins, proanthocyanidins, tannins, isothiocyanates, carotenoids, allyl sulfides, polyphenols, and phenolic acids. The difference between vitamins and these other phytochemicals is that deficiency states like scurvy have been defined for vitamins but not for the others. But make no mistake: These other phytochemicals are important, too. And Pauling's recommendation to ingest massive quantities of vitamins apart from their natural surroundings was an unnatural act. For example, as described in Catherine Price's book, *Vitamania,* half of an apple has the antioxidant activity of 1,500 milligrams of vitamin C, even though it contains only 5.7 milligrams of the vitamin. That's because the phytochemicals that surround vitamin C in apples enhance its effect. Then there's the plant goldenseal, which contains a powerful antibacterial substance called berberine. If you eat goldenseal, berberine isn't toxic. But if you purify berberine away from the other phytochemicals in goldenseal, and eat the same amount of berberine that was in the plant, it *is* toxic. Other phytochemicals in goldenseal protect against berberine's toxic effects. Another example would be the powerful antioxidant lycopene, which is present in tomatoes and used to hawk everything from ketchup to marinara sauce. Studies of rats with prostate cancer showed that tomato powder (which contains all of the phytochemicals found in tomatoes) could reduce the size of the tumors to a much greater extent than large quantities of purified lycopene. In short, Linus Pauling's appeal to all things natural was anything but.

PAULING'S ADVOCACY GAVE BIRTH TO a vitamin and supplement industry built on sand. Evidence for this can be found by walking into a GNC center—a wonderland of false hope. Rows and rows of megavitamins and dietary supplements promise healthier hearts, smaller prostates, lower cholesterol, improved memory, instant weight loss, lower stress, thicker hair, and better skin. All in a bottle. No one seems to be paying attention to the fact that vitamins and supplements are an unregulated industry. As a consequence, companies aren't required to support their claims of safety or effectiveness. Worse, the ingredients listed on the label might not reflect what's in the bottle. And we seem to be perfectly willing to ignore the fact that every week at least one of these supplements is pulled off the shelves after it was found to cause harm. Like the L-tryptophan disaster, an amino acid sold over the counter and found to cause a disease that affected 5,000 people and killed 28. Or the OxyElite Pro disaster, a weight-loss product that caused 50 people to suffer severe liver disease; one person died and three others needed lifesaving liver transplants. Or the Purity First disaster, a Connecticut company's vitamin preparations that were found to contain two powerful anabolic steroids, causing masculinizing symptoms in dozens of women in the Northeast.

LINUS PAULING'S LEGACY IS MIXED. He was the first to marry quantum physics with chemistry, the first to link the fields of molecular and evolutionary biology, and one of the few who stood up to McCarthyism and nuclear proliferation. But later in his life Linus Pauling was indistinguishable from the country fair hucksters and snake-oil salesmen of a century before—the father of a $32-billion-a-year vitamin and supplement industry. "Linus Pauling paid for his extraordinary gifts with his failure to appreciate where they rightfully ended," wrote historian Algis Valiunas. "One cannot but think what a marvelous legacy would have been his if he had just known when to quit."

How could a man who was so devoted to the rigor, hard work, and hard thinking required to achieve what he had achieved be at the same time unwilling to look critically at studies that consistently showed he was wrong, including those performed in his own institute? Sadly, Pauling's not the only one. Other brilliant, award-winning, internationally recognized scientists have also succumbed to hubris—with disastrous results.

Two of those scientists were associated with the AIDS epidemic.

On June 5, 1981, the Centers for Disease Control and Prevention (CDC) published a report of an unusual outbreak: Five previously healthy gay men in Los Angeles had developed a rare form of fungal pneumonia *(Pneumocystis carinii)* typically seen only in cancer victims or in people with severe immune deficiencies. These men, however, had all been previously healthy. The report also included the story of another seemingly unrelated cluster of gay men in New York and California who had developed a highly aggressive form of cancer called Kaposi's sarcoma.

One month later, the *New York Times* reported 41 more cases of Kaposi's sarcoma, again all in gay men. By the end of the year, another 270 similar cases in gay men were reported; 120 of these men had died. The press called it "the gay plague."

On September 24, 1982, CDC officials gave the disease a name: AIDS, for acquired immunodeficiency syndrome. Then, they set up a task force to find out what was causing it. Clues started to accumulate. On December 10, 1982, the CDC reported the first case of AIDS in an infant who had received a blood transfusion. The following week, the CDC reported another 22 cases of unusual infections in infants.

On January 7, 1983, the CDC reported the first cases of AIDS in women who had had sex with men who had AIDS. The following month, Robert Gallo, a researcher at the National Institutes of Health (NIH), predicted that an unusual virus called a retrovirus was causing AIDS. Gallo's prediction was surprising; up to this point, retroviruses had been considered to be benign viruses that didn't cause diseases in people. The CDC agreed with Gallo, believing that the disease was likely caused by a virus that was transmitted sexually or by exposure to blood. On May 20, 1983, Luc Montagnier found the cause: a virus he called LAV for lymphadenopathy-associated virus.

On April 23, 1984, Margaret Heckler, secretary of the U.S. Department of Health and Human Services, announced that Robert Gallo and his colleagues at NIH had also discovered the cause of AIDS: a virus they called HTLV-III for human T-cell lymphotropic virus. Heckler predicted that a vaccine to prevent AIDS would be available in the next two years. (That was more than 30 years ago.) Researchers soon realized that LAV and HTLV-III were the same virus, and settled on a third name: HIV for human immunodeficiency virus.

In 1985, the FDA licensed the first commercial test to screen blood and blood products for HIV. In 1987, the FDA licensed the first anti-HIV drug called AZT (zidovudine). By 1989, 100,000 people in the United States were infected with the AIDS virus.

Since those initial reports, much has been learned about HIV. Researchers have determined that HIV reproduces itself and eventually kills immune cells called helper T cells, the most important immune cells in the body. Helper T cells help other immune cells make antibodies or kill virus-infected cells. In addition to paralyzing the immune system, HIV constantly mutates during a single infection.

In essence, victims are infected with hundreds of different types of HIV, making it virtually impossible to neutralize the virus with antibodies or to make an effective vaccine. The good news is that as researchers better understood how HIV reproduces itself, they made highly active antiviral drugs. Although these drugs don't cure AIDS, they at least have changed the disease from one that was invariably fatal to a chronic infection.

Enter Peter Duesberg.

In March 1987, Duesberg published an article in the journal *Cancer Research,* claiming that AIDS *wasn't* caused by HIV—which he considered to be a harmless virus—but by long-term use of recreational drugs like heroin, cocaine, and amyl nitrate (poppers) by gay men. His hypothesis failed to consider babies or hemophiliacs who had received contaminated blood transfusions, homosexual men who didn't use recreational drugs, or women who had acquired the disease from AIDS sufferers. Normally, the research community would have dismissed this kind of poorly reasoned article as coming from a crank who had ventured far from his field. But Peter Duesberg was no crank. And viruses *were* his field. Trained in Germany, Duesberg was a full professor in the Department of Cell and Molecular Biology at the University of California at Berkeley, receiving full tenure when he was only 36 years old. His meteoric rise was because, in 1970, he became the first scientist to identify a specific viral gene that caused cancer. For this remarkable achievement, in 1986, Peter Duesberg was elected to the National Academy of Sciences. That same year, he received an Outstanding Investigator Research Grant from NIH and was made a Fogarty Scholar in Residence.

In the 1990s, as more research continued to implicate HIV as the cause of AIDS, Duesberg modified his hypothesis. Now, he claimed that, in Africa, malnutrition caused AIDS; in wealthy Africans, anti-HIV drugs caused the disease; and in hemophiliacs, some as yet unidentified contaminant in transfused blood was the problem. The event that was probably the hardest for Duesberg to explain occurred when three laboratory workers were inadvertently infected with a highly purified clone of HIV. None of the workers were gay, used recreational drugs, were hemophiliacs, were malnourished, or lived in Africa; one developed a severe form of AIDS. Duesberg said that because the other two workers didn't develop AIDS, HIV still hadn't been proven to be the cause. (Duesberg wasn't alone among famous scientists who had become AIDS deniers and conspiracy theorists. When evidence supporting the notion that HIV caused AIDS was clear, Nobel Prize–winning Kenyan ecologist Wangari Maathai said that scientists had created HIV in the laboratory for biological warfare. And Kary Mullis, who had won the Nobel Prize in chemistry for his discovery of the polymerase chain reaction [PCR], also stated that there was "no scientific proof" that HIV caused AIDS.)

Scientists eventually stopped listening to Peter Duesberg and his unsubstantiated rants. But Duesberg was not to be denied. In 2000, one year after taking over the presidency of South Africa from Nelson Mandela, Thabo Mbeki convened a Presidential AIDS Advisory Panel. He asked Peter Duesberg to head it. At the time, South Africa had more people living with HIV than any other country; 1 in 5 South African adults were infected with the virus. Mbeki, like Duesberg, believed that AIDS science was flawed and that anti-HIV drugs were poison, likening scientists to Nazi concentration camp doctors. Duesberg gave Mbeki the intellectual heft he needed to deny anti-HIV drugs to South Africans suffering from AIDS; as a consequence, more than 300,000 South Africans died needlessly from the disease.

Duesberg remains unrepentant. "I had all the students I wanted and I had all the lab space I needed," he said. "I got all the grants awarded. I was elected to the National Academy [of Sciences]. I became California Scientist of the Year. All my papers were published. I could do no wrong . . . until I started questioning the claim that HIV [was] the cause of AIDS. Then everything changed." But the problem with Peter Duesberg wasn't that he questioned the contention that HIV caused AIDS; it was that he continued to deny a mountain of scientific evidence showing that it did. Like Linus Pauling before him, Duesberg simply refused to believe that he could ever be wrong.

Peter Duesberg didn't limit his denial to HIV. He also didn't believe that human papillomavirus (HPV) caused cervical cancer—an association that Harald zur Hausen proved and for which he won the Nobel Prize in 2008.

RECENTLY, ANOTHER FORMER Nobel Prize winner has also swerved from science: Luc Montagnier, the French researcher who, along with Françoise Barré-Sinoussi, had won the Nobel Prize for his discovery that HIV caused AIDS. In 2010, two years after he won the Nobel Prize, Montagnier—like Linus Pauling and Peter Duesberg before him—made a series of embarrassing public declarations.

First, Montagnier said that DNA molecules could be teleported from one test tube to another (presumably, in a manner similar to the way people were teleported in the television series *Star Trek*).

Then, Montagnier claimed that homeopathy made sense. Homeopathy is based on the now disproved belief that if you dilute a substance to the point that not a single molecule remains, the water in which it was diluted will remember that the substance was there. "I can't say homeopathy is right in everything," said Montagnier. "What I can say now is that high dilutions are right. High dilutions of something are not nothing. They are water structures which mimic the original molecules." (Given that there is a limited amount of water on Earth, you don't want it to remember where it's been.)

Finally, Montagnier joined the long list of those claiming a cure for autism. Montagnier said that when he took the blood of patients with autism—and diluted it to the point that not a single molecule of the original blood remained—he could detect electromagnetic waves indicating the presence of bacterial DNA. Autism, it appeared, was a bacterial infection. And it wasn't just autism that was caused by bacteria. Alzheimer's disease, Parkinson's disease, multiple sclerosis, rheumatoid arthritis, and chronic fatigue syndrome were bacterial infections, too.

In 2011, at the age of 78, Luc Montagnier left France to head a new division at the Jiao Tong University in Shanghai, China. Intent on proving that he was right, Montagnier studied a group of 200 children with autism, finding "spectacular" results in those who had received antibiotics. "[We] have observed that a long-term therapy consisting of successive antibiotic treatments with accompanying medications induced in 60 percent of cases a significant improvement, sometimes even a complete resolution of symptoms," crowed Montagnier. "These children can now lead a normal school and family life!"

After Montagnier discovered what he believed was a cure for autism, he submitted a paper describing his findings to an obscure journal of which he was the senior editor; it was accepted for publication in three days. Then he traveled to Chicago to present his findings to a national festival of false promises known as Autism One. Taking the podium next to people who claimed that autism could be cured with hyperbaric oxygen treatments, bleach enemas, and chemical castration, Montagnier said that the only thing these children really needed was a prolonged course of antibiotics. Resistance from the mainstream medical community was, according to Montagnier, to be expected. Such is the life of a medical maverick. "In 1983, we were only a dozen or so people to believe that the virus we isolated was the cause of AIDS," said Montagnier. "I'm just interested in helping these children."

WHAT HAPPENED TO LINUS PAULING, Peter Duesberg, and Luc Montagnier—all brilliant scientists who had clearly lost their way—all wedded to theories that were completely and utterly disproved by scientific studies?

One possible explanation is that these men had been so right for so long that even in the face of strong opposition, they could never imagine being wrong. Another possible explanation is that there is a fine line between genius and madness. Or maybe they just wanted to make the next big splash—something that would again thrust them into the limelight. Whatever the reason, all three men did an enormous amount of harm: Pauling, because he had convinced people to take large quantities of vitamins and supplements that have only increased their risks of cancer and heart disease; Duesberg, because he indirectly caused hundreds of thousands of South African deaths from AIDS; and Montagnier, because he took advantage of parents' desperate desire to help their children by offering a medicine that could not possibly help and, therefore, could only hurt.

THE LESSON IN THE LINUS PAULING STORY can be found in the movie *The Wizard of Oz:* **Pay attention to the little man behind the curtain**.

When first encountered, the Wizard of Oz was exactly as advertised: great and powerful. His voice was booming; his manner, intimidating; and his head, large, green, and oddly cerebral. But the Wizard wasn't what he appeared to be. When Toto pulled back the curtain, the Wizard was just a rumpled old man with a high-pitched, irritatingly nasal voice. Exposed, the Wizard said, "Pay no attention to that little man behind the curtain." But Toto's revelation was impossible to ignore. Dorothy was appalled. "You're a very bad man," she said. "No, my dear," replied the Wizard. "I'm a very good man. I'm just a very bad wizard." In the end, the Wizard of Oz was successful because he was a good psychologist, not a good magician. The same applies to science: Don't be blinded by reputation. Every claim, independent of a scientist's reputation, should stand on a mountain of evidence. No one should get a free pass.

When Linus Pauling claimed that proteins folded in a certain manner or that sickle-cell hemoglobin had a different electrical charge, he had reams of biochemical data to prove it. But when he claimed that vitamins and supplements made you live longer, he had only the word of Irwin Stone, a man who had no scientific credentials, had never published a scientific paper in his life, and didn't have a shred of evidence to back his contention.

Pauling counted on the "Wizard of Oz" effect to promote his belief that vitamins and supplements were miracle drugs. He hoped that people would ignore the little man behind the curtain (his lack of data) and pay attention only to the booming voice that came with having won two Nobel Prizes. Similarly, Rachel Carson was seductive because she was a dynamic storyteller: the most trusted science writer in America. Like Linus Pauling, Russell Portenoy's claim that oxycodone could offer pain relief without addiction and Walter Freeman's claim that lobotomies could cure psychiatric illnesses were persuasive because both men were respected members of the medical and scientific community. They were convincing because they wore the badge of academic success, not because they had reproducible data showing that they were right. Finally, Peter Duesberg's claim that HIV wasn't the cause of AIDS or Luc Montagnier's claim that bacteria caused autism were granted full public hearings because both men were brilliant, highly acclaimed virologists. The point is that all scientists—no matter how accomplished or well known—should have unassailable data to support their claims, not just a compelling personality, an impressive shelf of awards, or a poetic writing style.

# CHAPTER 8

# LEARNING FROM THE PAST

Although it's always easy in retrospect, several lessons have evolved from science's darker past. Let's see what happens when we apply these lessons to a variety of modern-day inventions like e-cigarettes, preservatives, chemical resins, autism cures, cancer-screening programs, and genetically modified organisms (GMOs).

**1. It's all about the data.**

Truths emerge when studies performed by different scientists working in different environments using different methods find similar results. Ignoring these truths can have disastrous consequences.

On its surface, this lesson would appear to be easy to follow—just look at the data. The problem with data, however, is that there's far too much of it. Every day about 4,000 papers are published in the world's medical and scientific literature. As you would expect, the quality of these studies follows a bell-shaped curve: Some are excellent; some are awful; and most are more or less mediocre. So how is one supposed to separate good data from bad? One way would be to focus on the quality of the journals. But even that isn't perfect. Excellent, peer-reviewed journals have published claims that excess coffee drinking causes pancreatic cancer; that the measles-mumps-rubella (MMR) vaccine causes autism; and that nuclear fusion—when two small nuclei combine to release energy—can occur at room temperature in a glass of water (cold fusion). All of these observations were later shown to be wrong by other investigators. ("The trouble with the world is not that people know too little," wrote Mark Twain, "it's that they know so many things that ain't so.")

So, if you cannot completely trust observations published in excellent scientific journals, what can you trust?

The answer is that science stands on two pillars: one more reliable than the other. The first pillar is peer review; before a paper is published, experts in the field review it. The process, unfortunately, is flawed. Not all experts are equal and sometimes bad data slip through. The second pillar saves the day: reproducibility. If researchers publish a paper claiming something fantastic (like the MMR vaccine causes autism), subsequent studies will prove it to be correct or not. For example, almost immediately after publication of the claim that MMR caused autism, hundreds of investigators working in Europe, Canada, and the United States tried to reproduce it. They couldn't. Study after study involving hundreds of thousands of children and costing tens of millions of dollars found that those who had received the vaccine were at no greater risk of autism than those who hadn't. Good science had won out.

**2. Everything has a price; the only question is how big.**

Even the most dramatic, lifesaving, groundbreaking, universally acclaimed scientific and medical breakthroughs—like antibiotics and sanitation programs—have come with a price. Nothing, as it turns out, is exempt.

The first antibiotic, sulfa, was invented in the mid-1930s. Next came penicillin, which was mass-produced during WWII. Antibiotics have saved our lives. Without them, we would continue to die routinely from pneumonia, meningitis, and a variety of other potentially fatal bacterial infections. In part because of antibiotics, we live 30 years longer than we did a hundred years ago. But aside from the problem of creating antibiotic-resistant bacteria, one consequence of antibiotic use was utterly unpredictable.

In the last decade or so, researchers have been studying something called the microbiome—the population of bacteria that line the surface of our skin, intestines, nose, and throat. Recently, researchers found a role for these bacteria that was surprising. The numbers and types of bacteria that cover our body in part determine whether we will develop diabetes, asthma, allergies, and obesity. More surprising, if we alter the bacteria of infants with antibiotics, we *increase* the risk of these disorders. The lesson is clear. Use antibiotics when you need them, but overuse them at your own risk.

Antibiotics aren't the only dramatic scientific advance that has come with an unanticipated price. Even sanitation programs, which have decreased the incidence of food-borne and waterborne diseases like salmonella, shigella, *E. coli,* and hepatitis A, have unseen risks. Although the incidence of potentially fatal bacterial and viral infections has decreased in the developed world, the incidence of diseases like asthma and allergies has increased—a phenomenon that cannot be accounted for solely by industrialization. The reason for the unanticipated problem with sanitation can be found in an editorial in the *New England Journal of Medicine* titled "Eat Dirt." Children in developing world countries are, from the moment of birth, bombarded with germs; as a consequence, their intestines often contain parasites and toxin-producing bacteria that are rare in the developed world. Although these infections can cause malnutrition and even death, they make it far less likely that children in developing world countries will develop allergies and asthma. Researchers call this "the hygiene hypothesis."

The point being that there's always a price to pay. The challenge is figuring out whether a particular technology is worth the price. And we shouldn't grandfather in certain technologies just because they've been around for decades, or even centuries. All technologies should be constantly evaluated. Perhaps the best example is general anesthesia. Although anesthetics have been around for more than 150 years, only recently has it become clear that they can cause problems with attention and memory that can last for years. "No particular anesthetic has been exonerated," says Roderic G. Eckenhoff, a professor of anesthesiology at the University of Pennsylvania.

**3. Beware the zeitgeist.**

Three current technologies have been victims of the current culture: e-cigarettes, because no one likes the image of a teenager smoking, even if it's not actually smoke; GMOs, because the technology smacks of hubris, our attempt to alter the natural order; and bisphenol A (BPA), because it is a chemical resin that can leach out of plastic baby bottles. All three technologies have been the victims of scientific studies purporting harm. And all three have suffered at the hands of the media. Negative press, however, shouldn't blind us to the evidence.

First introduced into the United States in 2006, e-cigarettes are battery-powered vaporizers that deliver nicotine but don't contain tobacco. The liquid that is vaporized also contains propylene glycol, glycerol, and assorted candy or dessert flavorings (such as Belgian waffle and chocolate). Only a letter and a hyphen away from one of the most destructive products ever invented (cigarettes), e-cigarettes have been universally condemned by virtually every scientist, physician, and government official responsible for the public's health. And it's not hard to understand why.

First, nicotine is highly addictive and potentially harmful, especially to a developing fetus. It can also cause headaches, nausea, vomiting, dizziness, nervousness, and a rapid heartbeat. Although some brands of e-cigarettes don't contain nicotine, most do.

Next, Big Tobacco companies like Altria, Reynolds, and Imperial make e-cigarettes. Although company executives claim that e-cigarettes are an exit strategy for those trying to quit smoking, they haven't exactly earned the trust of the American public. In 2012, e-cigarette makers spent more than $18 million on magazine and television ads. Unlike ads promoting cigarettes, which were banned in 1971, companies are free to advertise e-cigarettes. As a result, e-cigarettes have become a $3.5-billion-a-year industry in the United States, with some predicting that sales will exceed those for conventional cigarettes by the mid-2020s.

Finally, reminiscent of the Joe Camel commercials, some advertisements for e-cigarettes are specifically designed to entice young people. When Julia Louis-Dreyfus was shown smoking an e-cigarette during the 2014 Golden Globe Awards, both Henry Waxman (D-CA) and Frank Pallone, Jr. (D-NJ), called the president of NBC to say that the actress was "sending the wrong message to kids about these products." Protests by people like Waxman and Pallone have fallen on deaf ears; e-cigarettes have become enormously popular among the young. In 2013, about 250,000 minors who had never smoked a cigarette had tried an e-cigarette. In 2014, an estimated

1.6 million middle and high school students in the United States experimented with them, a dramatic increase from the previous year. Indeed, more than 10 percent of high school students in the United States have tried an e-cigarette. It seems like only a matter of time before this tidal wave of children using e-cigarettes will become a flood of adults smoking cigarettes—and dying from lung cancer as a result. E-cigarettes appear poised to add to the 480,000 deaths and $300 billion in direct health care expenditures and productivity losses caused each year by cigarette smoking.

For all of these reasons, the American Cancer Society, the American Lung Association, the Centers for Disease Control and Prevention (CDC), the World Health Organization, and the American Academy of Pediatrics all strongly oppose e-cigarettes. And, when I first approached this subject, I assumed that I would end up agreeing with them, wholeheartedly. There was, however, one problem: the data.

Associated with the dramatic rise in e-cigarettes during the past five years, cigarette smoking has declined to historically low levels, including among the young. For example, the CDC reported that although the use of e-cigarettes had tripled between 2013 and 2014, the incidence of cigarette smoking had declined dramatically. In 2005, 20.9 percent of adults smoked cigarettes; by 2014, only 16.8 percent did, a 20 percent drop. Indeed, in 2014, the number of Americans smoking cigarettes fell below 40 million for the first time in 50 years. Further supporting the notion that e-cigarettes were replacing cigarettes, states that had banned the sale of e-cigarettes to minors witnessed an increase in cigarette smoking in that age group. And there is no denying that e-cigarettes are safer; unlike cigarettes, they don't produce tars that cause cancer or combustibles like carbon monoxide that cause heart disease. "People smoke for the nicotine but they die from the tar," said Michael Russell, a pioneer of nicotine-cessation treatments.

Maybe this is all coincidence. Maybe there are other reasons that cigarette smoking is declining that have nothing to do with the rise of e-cigarettes. But it's too early to condemn e-cigarettes as a gateway product to cigarette smoking when the opposite appears to be true. Time will tell. The point being that the cultural milieu that damns e-cigarettes is irrelevant; only the data are relevant. In August 2015, England's Department of Health recommended e-cigarettes as an effective way to stop smoking. Nine months later, in April 2016, the Royal College of Physicians, an organization of British doctors founded in 1518, supported the Department of Health's decision. British physicians had been influenced by a study in the United Kingdom showing that smokers who used e-cigarettes were much more likely to quit than those who used nicotine patches.

Like e-cigarettes, GMOs have also fallen victim to the zeitgeist.

GMOs are defined as any living organism that possesses "a novel combination of genetic material obtained through the use of modern biotechnology." The key phrase here is "modern biotechnology," because the truth is that we have been genetically modifying our environment since the beginning of recorded history. Using breeding or artificial selection, humans began to domesticate plants and animals around 12,000 B.C.—all for the purpose of selecting for certain genetic traits and all a precursor to modern genetic modification. Nonetheless, for environmentalists, no single act of hubris has been more terrifying than when scientists decided to recombine DNA in the laboratory to modify nature.

Today, the largest use of genetic bioengineering has been in food production. Genetic engineering has allowed crops to resist pests, tolerate extreme temperatures and environmental conditions, and be free of certain diseases. Genetically engineered crops have also been created to improve nutrient profiles, lengthen shelf lives, and resist herbicides. In the United States, 94 percent of soybeans, 96 percent of cotton, and 93 percent of corn are genetically modified; in the developing world, 54 percent of crops are genetically modified. The consequences, especially for farmers in the developing world, have been dramatic. GMO technology has reduced chemical pesticide use by 37 percent, increased crop yields by 22 percent, and increased profits for farmers by 68 percent. Although GMO seeds are more expensive, the cost is easily offset by reduced use of pesticides and higher yields.

Although many people fear that genetically modified foods might be more dangerous than other foods, careful scientific studies show they have no reason for concern. The American Association for the Advancement of Science and the National Academy of Sciences have both issued statements supporting the use of GMOs. Even the European Union, which has never been particularly supportive of GMOs, cannot ignore the science. In 2010, the European Commission issued the following statement: "The main conclusion to be drawn from the efforts of more than 130 research projects, covering a period of more than 25 years of research involving more than 500 independent research groups, is that biotechnology, and in particular GMOs, are not *per se* more risky than conventional plant breeding technologies."

Although the science is clear, the public remains concerned. A recent Gallup poll found that 48 percent of the American public believed that genetically modified foods posed a serious risk to consumers. Many of those polled wanted foods to contain GMO warning labels so they could know which ones to avoid. This poll showed that not only are we willing to ignore science, but we're also willing to ignore history. Due to selective breeding and cultivation, the crops we raise today "naturally" have little resemblance to their ancestors. From a practical standpoint, the farmer taking advantage of a chance mutation to cultivate a specific crop is indistinguishable from a choice to create the mutation ourselves. Both have the same mutation.

Genetic modification has also been used to make lifesaving medicines. Insulin used by diabetics, clotting proteins used by hemophiliacs, and human growth hormone used by children with short stature have all been made using genetic engineering technology. Previously, these products were obtained from pig pancreases, blood donors, and the pituitaries of dead people.

Yet those who oppose GMOs persist. Recently, the story of a tomato containing a fish gene made the rounds. The Frankensteinian image galvanized environmentalists to push harder to label GMO foods. Steven Novella, an assistant professor at Yale University School of Medicine, and the creator of the podcast, *The Skeptics Guide to the Universe,* summed it up best: "The real question here is not whether there is a GMO tomato with a fish gene, but who cares?" he wrote. "It's not as if eating fish genes is inherently risky—people eat *actual* fish. Furthermore, by some estimates, people share about 70 percent of their genes with fish. You have fish genes and every plant you have ever eaten has fish genes. Get over it!"

The GMO controversy reached its illogical end in 2015, when New York Assemblyman Thomas J. Abinanti introduced Bill 1706, banning all genetically modified vaccines. Not surprisingly, most vaccines are genetically modified. If not, then people would be injected with the "natural" bacteria or viruses that caused the disease. For example, by genetically modifying poliovirus, we've eliminated polio from the United States and from much of the world. Vaccines have to be genetically modified.

Perhaps no single chemical has suffered from the zeitgeist more than bisphenol A (BPA).

In 1935, the DuPont chemical company launched its slogan, "Better Living Through Chemistry." In 1982, DuPont dropped "Through Chemistry" and later abandoned the slogan altogether in favor of "The Miracles of Science." The word "chemistry" just didn't sit well with the American public. We seem to respond negatively to anything with a chemical name. And bisphenol A certainly fits that bill.

BPA, which was first synthesized in 1891, wasn't commercially available in the United States until 1957, when it was used to make plastics and resins. The chemical found its way into goggles, face shields, bicycle helmets, water bottles, baby bottles, CDs, DVDs, the lining of water pipes, and the lining of metal soup and soda cans. Ironically, BPA wasn't invented to make plastic clear and tough. It was invented as a synthetic estrogen, the hormone primarily responsible for regulating the female reproductive system. But BPA was a weak estrogen—about 40,000 times weaker than other synthetic estrogens—so it was abandoned, only later to be picked up as a plasticizer. What researchers soon discovered, however, was that this weak hormone, although insoluble in water, could leach out of plastic or metal containers. They feared that Americans, including American babies, might unknowingly be ingesting a feminizing hormone.

Concerned that BPA might be harmful, researchers studied its effects on mice and rats, linking it to breast cancer, prostate cancer, early onset puberty, ovarian cysts, obesity, and even attention deficit disorder. Then they started investigating people, finding that 93 percent of adults had traces of BPA in their urine. "If you don't have BPA in your body," wrote one *Time* magazine reporter, "you're not living in the modern world."

Armed with this information, Nalgene, which makes plastic containers, removed BPA from all of its products. Then the FDA banned its use in baby bottles. BPA was now, according to one reporter, "among the world's most vilified chemicals." Similar to the story of DDT, however, the BPA story also soon fell apart.

Initially, researchers had trouble reproducing the animal model studies, especially when using quantities of BPA likely to be encountered by people. A 2004 report from the Harvard Center for Risk Analysis found "no consistent affirmative evidence for low-dose BPA effects." Glenn Sipes, who was a co-author of the study, said, "I've never had a problem saying that we can see biological effects in these low doses. But why are we seeing these studies that can't be repeated? Why do we have to work so hard to try and replicate and show these low doses really have an effect? Why don't [problems with BPA] stand out in black and white?"

In 2011, a review of studies in people found no evidence that low doses of BPA caused harm. The reason that studies in rodents had found that BPA had caused problems was that the rodents had been *injected* with BPA; injection bypassed the liver, which typically inactivates BPA within five minutes. When rodents were *fed* BPA instead of being *injected* with it, those given 40, 400, or 4,000 times the typical human exposure remained healthy.

In July 2014, the FDA stated that, "BPA is safe at the current levels occurring in foods." Similarly, the European Food Safety Authority, which published recommendations about BPA in 2008, 2009, 2010, 2011, and 2015, also continues to state that BPA is safe. Both agencies have set limits for the tolerable daily intake (TDI) for BPA. To exceed this limit, an average adult would have to ingest about 10,000 times more BPA than one would typically ingest in a day: the equivalent of eating more than 500 cans of soup. Nonetheless, today it's hard to find a water bottle that doesn't proudly proclaim "BPA-free" on the label.

Another reason that we should have been suspicious of the BPA studies is that mice aren't men. All studies of experimental animals should be viewed with caution. For example, in the early 1970s, saccharin was shown to cause bladder cancer in rodents. As a result, all food containing saccharin bore a label warning of its dangers. By 2000, scientists realized that what was happening in rodents wasn't happening in people. The reason was that, unlike humans, the urine of rodents is highly acidic and contains large quantities of calcium phosphate and proteins. For these reasons, rodents fed saccharin formed microcrystals in their urine that damaged the lining of the bladder, causing bladder cancer. None of these events occurred in people. On December 21, 2000, the FDA removed warning labels from foods containing saccharin.

Also, if you're going to say that animal studies predict events in people, then we should stop eating chocolate, which can cause heart arrhythmias and occasionally death in dogs. As it turns out, dogs cannot tolerate even small amounts of a substance in chocolate called theobromine. People, on the other hand, can consume much larger quantities of chocolate without getting sick. (I am living proof of this.)

Animal studies can also be misleading for another reason: They can show that something is valuable even when it isn't. For example, early studies of a vaccine to prevent HIV were promising in experimental mice and monkeys. But studies in people have been far less promising. "Mice lie and monkeys exaggerate," says University of Pennsylvania vaccine researcher David Weiner.

Our fear of anything with a chemical name isn't likely to go away anytime soon. A few years ago the comedians Penn and Teller did an experiment. They sent a friend to a fair in California to get signatures on a petition to ban dihydroxymonoxide. Hundreds of people signed the petition, convinced that the chemical was bad for you. *Dihydrox* means two hydrogen (H) atoms and *monoxide* means one oxygen (O) atom. The combination, $H_2O$, is water. By using a chemical name, their friend was able to convince hundreds of people to ban water from the face of the Earth.

**4. Beware the quick fix.**

Like lobotomies, mental institutions bursting at the seams with adult schizophrenia patients are a relic of the past. Schizophrenia has become an outpatient disease. So has autism, arguably the most common psychiatric disorder of children. As a consequence, the pressures to find a cure have shifted from psychiatrists working at public facilities to parents living in private homes. Unfortunately, like their counterparts in the past, parents have become desperate, willing to do anything to relieve the suffering. So, although we may think that gruesome, ill-conceived, medieval therapies like lobotomies are behind us, they're not.

Children with autism have been put in hyperbaric oxygen chambers, causing intense, painful pressure on their eardrums and, in one case, death. They've been given intravenous medicines designed to bind heavy metals, causing another child to die when his heart stopped beating. They've been taken to Mexico or other countries for stem cell transplantations. And, perhaps worst of all, they've been subjected to a therapy invented by a former Scientologist turned health evangelist named Jim Humble, who calls himself the archbishop of the Genesis II Church of Health. In his online video, Humble claims to be a billion-year-old god from the Andromeda galaxy.

Humble believes that autism—as well as AIDS, malaria, cancer, and Alzheimer's disease—is caused by worms living in the intestine.

To kill the worms, he invented what he calls the Miracle Mineral Solution or MMS. MMS contains sodium chlorite and citric acid, which combine to form chlorine dioxide, a powerful bleach. MMS, which children swallow or receive as an enema, is now quite popular in the autism community. The problem, apart from the fact that autism isn't caused by worms, is that even small quantities of MMS can cause nausea, vomiting, diarrhea, intestinal bleeding, respiratory failure, hemolysis (when red blood cells in the bloodstream break apart), and, ironically, developmental delay. In October 2015, one U.S. vendor was sent to prison for selling the product. MMS has been linked to at least one death.

Parents who subject their children to MMS often share their stories online. They write about children crying out in pain. They show pictures of the lining of their children's intestines that have come out in their stools, believing, wrongly, that they're worms. They talk about how their children's hair has fallen out. And they talk about how their children have slowly grown more apathetic, losing any previous emotion. How, as they have chronically poisoned their children with an industrial bleach, their children have quieted down, becoming much easier to handle. In essence, how—as had been the case for lobotomies—they have merely substituted one disorder for another. Still, these parents urge each other on. It's working, they claim.

The contrasts between lobotomy and MMS therapy are striking. Lobotomies were endorsed by the American Medical Association, the American Psychiatric Association, and the *New England Journal of Medicine*. MMS therapy has never been endorsed by any professional or medical organization; on the contrary, the FDA has issued a warning against its use. Ice pick lobotomies were invented by a respected neurologist who was a professor at a well-known medical school. MMS therapy was invented by a man claiming to be from a galaxy

2.5 billion light-years from Earth. Frankly, it's much easier to understand how people could lobotomize their children than squirt a powerful industrial bleach into their children's mouths and rectums.

Although it might seem far-fetched, imagine the following scenario: A group of unscrupulous doctors opens a clinic in Switzerland that performs lobotomies for the treatment of autism. The doctors who run the clinic don't call the procedure a lobotomy (that ship has sailed); they call it something else, like the "Fresh Start" procedure. The doctors put up an attractive website that explains how the outpatient procedure takes only a few minutes to perform and involves severing the nerve fibers in the brain that cause autism. They include a few parent testimonials stating that after the procedure their children's language doubled, and then they open their doors. If a man claiming to be a billion-year-old god from another galaxy can get people to line up for a therapy that crosses the line into child abuse, then doctors in some European clinic should expect at least equal success with a quick-fix surgical procedure. It hasn't happened yet, but there is no reason to believe that it couldn't. Desperate to do something, anything, to cure the incurable, we continue to punish the afflicted.

**5. The dose makes the poison.**

When Rachel Carson wrote *Silent Spring*, she correctly predicted that man-made activities could destroy the environment. Thanks to Rachel Carson, we are now far more attentive to our impact on the planet. Unfortunately, Carson also gave birth to the notion of zero tolerance—the assumption that any substance found harmful at any concentration or dosage should be banned absolutely. If large quantities of DDT (like those used in agriculture) were potentially harmful, then even small quantities (like those used to prevent mosquitoes from biting) should be avoided. As a result, millions of children died needlessly from malaria.

One recent example of how the concept of zero tolerance has caused harm is thimerosal: an ethylmercury-containing preservative used in vaccines. Because vaccines are injected into babies, no preservative has generated more angst.

Preservatives were first added to multidose vials of vaccines back in the 1930s. The reason was that as the needle penetrated the rubber stopper again and again, bacteria could enter the vial. In the early 1900s, children receiving the eighth, ninth, or tenth dose of a vaccine might be inoculated with bacteria that had inadvertently been introduced into the vial, causing severe and occasionally fatal infections. By adding preservatives containing ethylmercury to vaccines, this problem disappeared. But in the late 1990s the question arose, at what price?

In 1999, some physicians worried that babies and young children might be exposed to too much mercury in vaccines. What happened to thimerosal in vaccines in the 1990s eventually mimicked what had happened to DDT in the 1970s. Exercising caution, thimerosal was precipitously removed from infant vaccines and branded with a scarlet letter. During the year or so when manufacturers were gearing up to produce thimerosal-free vaccines, some, like the hepatitis B vaccine, still contained thimerosal. About 10 percent of hospitals decided not to give thimerosal-containing hepatitis B vaccines; as a result, one three-month-old child in Michigan died of overwhelming hepatitis B virus infection and six children in Philadelphia born to mothers infected with the virus weren't vaccinated, condemning them to the likely development of chronic liver disease (called cirrhosis) or liver cancer later in life. These hospitals had wrongly assumed that the harm from a thimerosal-containing hepatitis B vaccine (which was theoretical at best) was greater than the risk of getting hepatitis B (which wasn't theoretical at all). Within a few years of the removal of thimerosal from vaccines given to young children, seven studies showed that it hadn't caused harm. The only harm had come from elevating a theoretical risk above a real risk.

It's not hard to understand how this could have happened. Mercury is never going to sound good. (There's no such thing as the National Association for the Appreciation of Heavy Metals.) Large quantities of mercury have clearly been shown to cause harm. Indeed, environmental mercury contaminations caused by a chemical spillage in Minamata Bay, Japan, or by fumigating grain in Iraq damaged hundreds of babies and fetuses. Mercury, however, which is present in the Earth's crust, is hard to avoid. Anything we drink that's made from water on this planet (including breast milk and infant formula) will contain small quantities of mercury. These small quantities of mercury aren't harmful. Only large quantities are harmful. If small quantities of mercury were harmful, we'd have to move to a different planet. Of interest, the quantities of mercury contained in vaccines were far less than those contained in breast milk and infant formulas. And mercury isn't the only heavy metal to which we are routinely exposed. We all have trace quantities of metals like cadmium, beryllium, thallium, and even arsenic in our blood. But at quantities far less than those required to do harm.

Unfortunately, we seem incapable of learning the most important lesson in toxicology: The dose makes the poison. For example, if people drink several gallons of water quickly (as has occurred during fraternity hazings), they might exceed their body's ability to hold onto the mineral sodium. As the level of sodium in the bloodstream drops, it's possible that they could have a seizure. This doesn't mean that water is toxic to the brain. It only means that you shouldn't drink several gallons of it at once. Similarly, a cup of organic coffee contains acetaldehyde, benzaldehyde, benzene, benzopyrene, benzofuran, caffeic acid, catechol, 1,2,5,6 dibenzanthracene, ethylbenzene, formaldehyde, furan, furfural, hydroquinone, d-limonene, 4-methylcatechol, styrene, and toluene. Many of these substances are known cancer-causing agents or DNA-altering chemicals. Yet no study has ever found that organic coffee causes cancer. That's because the quantities of the chemicals contained in coffee are well within the levels considered safe. Rachel Carson's lesson of zero tolerance doesn't apply to the real world.

**6. Be cautious about being cautious.**

Rachel Carson taught us caution. Wouldn't it make sense to ban DDT given that at the very least, it might cause harm to people. As we learned, removing it from use caused far more harm than good.

However, one could make a reasonable argument that, regarding BPA, what possible harm could come from removing a plasticizer from a toy. At the time that the FDA banned it from baby bottles and Nalgene removed it from all of its products, it wasn't clear whether BPA was safe. Wouldn't it make sense, then, to remove BPA in the interest of caution—under the precautionary principle? The answer is yes; it was reasonable. But we have to be careful. We have to make sure that in the name of the precautionary principle we aren't doing more harm than good. Although removing a plasticizer from a child's toy was without consequence (and could have been predicted to have been without consequence), the decisions to precipitously remove thimerosal from vaccines (and scaring parents and doctors) or to ban DDT from public health programs (when the only replacements were more expensive and less available) have caused children to suffer needlessly and predictably. The precautionary principle, at the very least, assumes that no harm will come from exercising caution.

Which brings us to arguably the single greatest use of the precautionary principle in modern medicine: cancer-screening programs.

During the past 50 years, doctors and scientists have proven that some cancers can be prevented. Sunblock can prevent skin cancer. The hepatitis B virus vaccine can prevent the most common cause of liver cancer. The human papillomavirus vaccine can prevent the only known cause of cervical cancer. And cessation of cigarette smoking can prevent a common cause of lung cancer. The results of these four strategies are clear.

The definition of cancer, however, is changing—and not for the better. Medical textbooks 20 years ago defined cancer as a "disease the natural course of which is fatal." No longer. Now we detect cancers that aren't fatal—the kind of cancers that you are more likely to die *with* than *from*. In the process of detecting these nonfatal cancers, we are probably doing more harm than good.

Gilbert Welch, a professor at Dartmouth Medical School, offers the best analogy to our current dilemma. It's a barnyard analogy. Imagine, writes Welch, three animals in a barn that are trying to escape: a bird, a turtle, and a rabbit. When you open the door, they will escape at different rates. The bird, which will invariably fly out before you have time to close the door, is analogous to the kind of cancer that will kill you no matter what you do. Even if you detect the cancer early, it doesn't matter—you'll die from it anyway. It's just too aggressive. The turtle, which is so plodding that it will never effectively escape, is analogous to a cancer that is so slow growing, so nonvirulent, that it will never kill you. You will invariably die from something else first. This is the kind of cancer that you will die with and not from. The rabbit, which can be caught if the door is closed quickly enough, is analogous to the kind of cancer worth detecting. If this cancer isn't detected early, then it will kill you.

And if it is detected early, then the screening test will have saved your life.

Screening tests are valuable only if they are detecting rabbits; if they're detecting mostly turtles and birds, then they won't save lives. Some screening tests, like Pap smears to detect cervical cancer or colonoscopies to detect colon cancer, save lives. Both of these tests are detecting lots of rabbits. For thyroid, prostate, and breast cancers, on the other hand, the value of early screening tests isn't so clear. Atul Gawande, a surgeon at Johns Hopkins University School of Medicine and prolific author, says it best: "We now have a vast and costly health care industry devoted to finding and responding to turtles."

We'll start with thyroid cancer.

In 1999, the South Korean government initiated a large nationwide screening program for the early detection of thyroid cancer. The screening test used was ultrasonography, which sends high-frequency sound waves (higher than those audible to the human ear) into the body. The sound waves then bounce back. Different structures absorb or reflect sound waves differently. As a result of the massive screening program, South Korean physicians detected more than 40,000 new cancers of the thyroid, 15 times more than had been detected before the screening program started. Thyroid cancer became the most common cancer in South Korea. One researcher called it a "tsunami of thyroid cancer."

Virtually all of these South Korean thyroid cancers were treated by completely removing the thyroid, called a thyroidectomy. This procedure, however, comes with a price. At the very least, everyone with a thyroidectomy has to take thyroid replacement hormones for the rest of their lives. And sometimes it's difficult to get the dose right. People suffer from symptoms of too much replacement hormone (sweating, heart palpitations, and weight loss) or too little replacement hormone (somnolence, depression, and weight gain). Worse, because the nerves of the vocal cord travel close to the thyroid gland, some people suffer vocal cord paralysis. Or they suffer from a problem with calcium metabolism because a gland called the parathyroid gland, which regulates calcium, is also nearby. Or they suffer from life-threatening bleeding after the surgery. Initially, South Korean health officials were excited that they had detected all of these cancers before patients had developed any symptoms. Then they looked at the mortality rates from thyroid cancer. No difference. The incidence of death from thyroid cancer was exactly the same before and after the massive screening program. The only tangible outcome was that now tens of thousands of South Koreans had to suffer the side effects of thyroid surgery.

Overdiagnosis and overtreatment of thyroid cancer hasn't been limited to South Korea. In France, Italy, Croatia, Israel, China, Australia, Canada, and the Czech Republic, the rates of thyroid cancer have more than doubled. In the United States, they've tripled. In all of these countries, as had been the case in South Korea, the incidence of death from thyroid cancer has remained the same.

Autopsy studies confirm the problem with trying to screen for thyroid cancer. About one-third of people who have died from other causes also had thyroid cancer at autopsy. Some researchers argue that if the sections of the thyroid examined in these studies had been thinner and more numerous, the percentage of people with thyroid cancer at the time of their death would have approached 100 percent. Which isn't to say that people don't die from thyroid cancer. They do. The death rate from thyroid cancer in the United States is about 1 per 200,000 people. The problem with thyroid cancer is that it's almost all turtles and a few birds. There just aren't enough rabbits to make screening worthwhile.

Next year in the United States, about 60,000 people will be diagnosed with thyroid cancer. Women will outnumber men three to one. Virtually all will have thyroidectomies, and few, if any, will benefit from the diagnosis. If most of these small thyroid cancers don't kill you, then maybe we shouldn't call them cancer.

Screening for prostate cancer has also come under closer scrutiny.

In 1970, Richard Ablin, a professor of pathology at the University of Arizona, discovered the PSA test, which stands for prostate-specific antigen. PSA is an enzyme made by cells in the prostate gland. The purpose of PSA is to break up cervical mucus so that sperm can enter the uterus. Criminologists were the first to recognize the value of PSA, which offered proof for the existence of semen in rape cases, even when the rapist had had a vasectomy or couldn't make sperm. Doctors were the next to realize the potential value of PSAs, using the test to determine whether prostate cancers had relapsed. Then physicians took a step they are now starting to regret—PSA tests were used to *predict* whether someone had prostate cancer. If the level of PSA in the blood was high, then urologists recommended a prostate biopsy. If the biopsy showed prostate cancer, then men either had a total removal of the prostate (prostatectomy) or they had radiation therapy to the prostate. More than 90 percent of men in the United States who have been diagnosed with prostate cancer have had one therapy or the other.

Because of the PSA test, prostate cancer is now the most commonly diagnosed non-skin cancer in the United States. So, what's happened to the incidence of death from the disease? Nothing. The risk of dying from prostate cancer hasn't changed in the last ten years. Indeed, about 50 percent of men more than 60 years old have been found at autopsy to have prostate cancer after they had died from something else; in men more than 85 years old, that number climbs to 75 percent. In other words, as had been the case for thyroid cancer, men are more likely to die with prostate cancer than from it. Prostate cancer, like thyroid cancer, is mostly turtles and birds.

In 2012, the United States Preventive Services Task Force recommended against PSA-based screening tests for prostate cancer. But not before a lot of harm had been done. High PSA tests invariably lead to prostate biopsies, which can cause pain, bleeding, difficulties with urination, and bloodstream infections requiring hospitalization. And, aside from the psychological trauma of being diagnosed with prostate cancer, therapies for the disease are brutal. Prostate surgeries and radiation commonly cause incontinence and erectile dysfunction. Worse, five of every thousand men will die from prostate surgery. And all for nothing. By 2015, three years after the task force had issued its recommendations, the number of people screened for and diagnosed with prostate cancer had declined; many physicians had gotten the message.

Two years before the task force changed its recommendation, Richard Ablin, the discoverer of PSA, penned an op-ed piece for the *New York Times*. Noting that the annual bill for PSAs was $3 billion, he wrote, "As I've been trying to make clear for many years now, PSA testing can't detect prostate cancer and, more important, it can't distinguish between the two types of prostate cancer—the one that will kill you and the one that won't. I never dreamed that my discovery four decades ago would lead to such a profit-driven public health disaster."

Mammography screening for breast cancer is also being reevaluated. Although it is clear that mammography, which was first introduced into the United States in the mid-1970s, saves lives; the question is how many and at what cost.

In 2012, Archie Bleyer and Gilbert Welch published a study in the *New England Journal of Medicine* titled "Effect of Three Decades of Screening Mammography on Breast-Cancer Incidence." They found that, with the advent of screening mammographies, the incidence of breast cancer in the United States had doubled. For every 100,000 women screened, the number of women diagnosed with breast cancer had increased from 112 to 234. In other words, every year 122 more women (per 100,000) were being diagnosed with breast cancer. At the same time, the number of women presenting with late-stage breast cancer (the kind that often results in death), decreased from 102 to 94 (per 100,000). This meant that only 8 women among 122 appeared to benefit from the screening. Eight. The others were treated with mastectomy, radiation therapy, and chemotherapy without benefit. The authors concluded that, although the incidence of death from breast cancer had clearly decreased during the era of screening mammography, most of that decrease was due to better treatment, not better screening. They also estimated that during the three decades of mammography, about 1.3 million women had been diagnosed with a cancer that would never have killed them. The authors concluded that, "screening is having, at best, only a small effect on the rate of death from breast cancer."

Another study involving hundreds of countries also cast doubt on the time-honored notion that mammographies save lives. Investigators found that different countries had different rates of screening. In some countries, as few as 40 percent of women were screened; in others, as many as 80 percent were screened. If mammographies were making a difference, then countries that screened more women should have had lower rates of deaths from breast cancer. But deaths rates from breast cancer were the same in both groups of countries. The only difference was that in countries with higher rates of screening, more women had mastectomies, radiation, and chemotherapy.

with no obvious benefit. As a consequence of these and other studies, the recommendations for mammography have changed. Doctors used to recommend that all women between 40 and 74 years of age get a mammogram every two years. Now the United States Preventive Services Task Force recommends screening starting at 50, not 40 years of age. There were just too many noncancers that were called cancers—and women were suffering needlessly as a result.

In February 2015, Christie Aschwanden, a journalist, wrote an editorial in the *Journal of the American Medical Association.* The article was titled, "Why I'm Opting Out of Mammography." Aschwanden imagined five possible outcomes following the procedure: One, "most likely the scan would turn up nothing suspicious." Two, "I'd be called back for further testing, perhaps even a biopsy for something that was not cancer," resulting in "sleepless nights [and] some lingering worry afterwards." Three, "the mammogram would find a cancer that would have remained innocuous if not detected. If a mammogram found one of these cancers—and right now it's impossible to definitively differentiate them from the dangerous ones—I would be treated and 'cured' for a cancer that was never destined to hurt me." Four, "the mammogram could find a very aggressive, incurable cancer—the kind responsible for most deaths. In this case, I might be diagnosed sooner, but I'd ultimately die from the cancer anyway, and I'd spend more of the years I had left receiving cancer treatment." Five, "the mammogram could find a dangerous cancer that's amenable to treatment, and my life would be saved." Using data from a recent study, Aschwanden reasoned that the chance that a mammogram would save her life was about 0.16 percent. She concluded that "a mammogram is more likely to 'cure' me of a harmless cancer—by subjecting me to life-disrupting and potentially harmful treatments like chemotherapy and radiation—

than it is to prevent me from dying of breast cancer. For me that's a deal breaker."

Until scientists are able to find genetic or biochemical markers that clearly distinguish bad cancers from innocuous ones, we are going to continue to suffer from the overdiagnosis and overtreatment of cancers that aren't really cancers at all. And continue to be told that our lives are being saved when they're not. Today, about 70,000 women every year are diagnosed with breast cancers that would never have killed them. Our overzealous sense of caution has caused a lot of unnecessary fear, anxiety, and debilitating surgeries.

**7. Pay attention to the little man behind the curtain.**

Today, it's not hard to find people who give medical or scientific advice based on the Wizard of Oz effect. Health gurus all hope that their winning personalities will hide their lack of evidence. And they don't like to be challenged. When little men behind curtains are revealed to be just little men behind curtains, they often cry foul. It wasn't that their claims were wrong, they argue, it was that evil forces were conspiring to defeat them.

For example, in 1998 a British physician named Andrew Wakefield proposed that the combination measles-mumps-rubella (MMR) vaccine caused autism. Thousands of parents in the United Kingdom and the United States withheld the MMR vaccine. As a consequence, hundreds of children were hospitalized and at least four died from measles. The public health and academic communities responded by performing more than a dozen studies. Their findings were clear, consistent, and reproducible. The MMR vaccine didn't cause autism. Andrew Wakefield had been wrong.

Were Wakefield a real scientist—which is to say someone who was open to the possibility that his hypothesis might be wrong—he would have looked at the mountain of evidence in front of him and stepped aside. But he didn't. Wakefield, as it turned out, possessed something that no scientist should ever possess: a nonfalsifiable hypothesis. MMR caused autism and that was that. So he did what all pseudoscientists do when they are shown to be wrong: He found a bogeyman, claiming that other researchers couldn't reproduce his findings because they had all been unduly influenced by the pharmaceutical industry. Wakefield wanted us believe that thousands of researchers, public health officials, academicians, and pediatricians on several continents were all deeply, hopelessly in the pockets of drug companies. Free of such undue influence, he implied, his theory was unassailable.

Linus Pauling was no different. When his hypothesis that vitamin C could treat cancer was refuted by two excellent studies published in the *New England Journal of Medicine,* he threatened to sue the journal. According to Pauling, the problem wasn't that he was wrong (after all, he was Linus Pauling, the holder of two Nobel Prizes), it was that there was a conspiracy by the medical establishment to defeat him. A medical establishment that had no interest in allowing a product as inexpensive as vitamin C to replace the expensive chemotherapies it had been cashing in on for decades.

If Andrew Wakefield and Linus Pauling's hypotheses were right, then subsequent studies would have shown that they were right. When well-designed studies refuted their claims, they chose to attack those who had found them to be wrong. They did what any good lawyer would do; they argued conspiracy. (The legal aphorism is that when the law is on your side, argue the law; when the facts are on your side, argue the facts; when neither is on your side, attack the witness.) The minute that you hear researchers claim conspiracy, you should suspect that their hypotheses are built on sand. And although their this-is-what-happens-when-you-speak-truth-to-power lament is appealing, it doesn't mean that it's right. As Norman Levitt, a mathematician and debunker of pseudoscience, famously said, "While Galileo was a rebel, not all rebels are Galileo"—no matter how hard they try to convince you that they are.

# EPILOGUE

*"History is the error we are forever correcting."*

—Anthony Marra, *The Tsar of Love and Techno*

When Pandora's curiosity got the best of her, she opened the forbidden box, unknowingly allowing hunger, pestilence, sickness, poverty, crime, and vice to escape into the world. Only one thing remained—hope. When Pandora opened the box again, hope also entered the world, with a lot of catching up to do.

Today, although the terms have changed, the concepts remain the same. Now the evils released from Pandora's box have more specific names like pests, vermin, bacteria, viruses, fungi, parasites, toxins, cancer, heart disease, and pain—all of which have inflicted suffering or limited lives. And so we fought back; our hope for a better world provided by scientific and medical advances. But our choice to push back against these evils means that we have chosen to engage in a kind of war. And, as in all wars, there have been casualties. Every advance has come with a price. Our task has been to figure out whether the price is too high. Sometimes, as has been the case for vaccines, antibiotics, and sanitation programs, the price has been small. Other times, such as for trans fats, lobotomies, and megavitamins, the price has been great. In each of these cases, the calculations have been easy. Many times, however, as has been the case for opiates and synthetic fertilizers, the calculations have been much tougher, where gains in the short term might be overwhelmed by losses in the long term.

In the end, although we hold on to the hope of a better life through science, we need to approach all scientific advances cautiously and with eyes wide open—and to make sure that we learn from our mistakes and aren't simply paralyzed by them.

# ACKNOWLEDGMENTS

I want to thank Susan Tyler Hitchcock at National Geographic for her patience and steady hand in guiding me through this book, and Louis Bell, Jeffrey Bergelson, David Gorski, Charlotte Moser, Brian Fisher, Will Offit, Bonnie Offit, Sally Satel, and Laura Vella for their careful readings of the manuscript.

# SELECTED BIBLIOGRAPHY

## GENERAL

Grant, John. *Discarded Science: Ideas That Seemed Good at the Time.* Surrey, England: Facts, Figures & Fun, 2006.

Grant, John. *Corrupted Science: Fraud, Ideology, and Politics in Science.* Surrey, England: Facts, Figures & Fun, 2008.

Grant, John. *Bogus Science: Or, Some People Really Believe These Things.* Surrey, England: Facts, Figures & Fun, 2009.

Livio, M. *Brilliant Blunders: From Darwin to Einstein—Colossal Mistakes by Great Scientists That Changed Our Understanding of Life and the Universe.* New York: Simon & Schuster, 2013.

## GOD'S OWN MEDICINE

Adams, Taite. *Opiate Addiction: The Painkiller Addiction Epidemic, Heroin Addiction and the Way Out.* Petersburg, Florida: Rapid Response Press, 2013.

Anonymous. "Closing Arguments Made in Trial of Doctor in Oxy-Contin Deaths," *New York Times,* February 19, 2002.

Anonymous. "Doctor Given Long Prison Term for 4 Deaths Tied to OxyContin," *New York Times,* March 23, 2002.

Ballantyne, J. C., and J. Mao. "Opioid Therapy for Chronic Pain," *New England Journal of Medicine* 349 (2003): 1943–1953.

Belluck, P. "Methadone, Once the Way Out, Suddenly Grows as a Killer Drug," *New York Times,* February 9, 2003.

Booth, Martin. *Opium: A History*. London: Simon & Schuster, 1996.

Brownstein, M. J. "A Brief History of Opiates, Opioid Peptides, and Opioid Receptors," *Proceedings of the National Academy of Sciences* 90 (1993): 5391–5393.

Califf, R. M., J. Woodcock, and S. Ostroff, "A Proactive Response to Prescription Opioid Abuse," *New England Journal of Medicine* 374 (2016): 1480–1485.

Carise, D., K. L. Dugosh, A. T. McLellan, et al. "Prescription OxyContin Abuse Among Patients Entering Addiction Treatment," *American Journal of Psychiatry* 164 (2007): 1750–1756.

Catan, Thomas, and Evan Perez. "A Pain-Drug Champion Has Second Thoughts," *Wall Street Journal,* December 17, 2012.

Centers for Disease Control and Prevention. "Prescription Painkiller Overdoses in the US," www.cdc.gov/vitalsigns/painkilleroverdoses, 2012.

Centers for Disease Control and Prevention. "CDC Guideline for Prescribing Opioids for Chronic Pain—United States, 2016," *Morbidity and Mortality Weekly Report* 65 (2016): 1–49.

Cicero, T. J., M. S. Ellis, and H. L. Surratt. "Effect of Abuse-Deterrent Formulation of Oxycontin," *New England Journal of Medicine* 367 (2012): 187–189.

Clines, F. X., and B. Meier. "Cancer Painkillers Pose New Abuse Threat," *New York Times,* February 9, 2001.

Courtwright, David T. *Dark Paradise: A History of Opiate Addiction in America.* Cambridge: Harvard University Press, 2001.

Courtwright, D. T. "Preventing and Treating Narcotic Addiction—A Century of Federal Drug Control," *New England Journal of Medicine* 373 (2015): 2095–2097.

Dormandy, Thomas. *Opium: Reality's Dark Dream*. New Haven: Yale University Press, 2012.

Fernandez, Humberto, and Therissa A. Libby. *Heroin: Its History, Pharmacology, and Treatment*. Center City, Minnesota: Hazelden, 2011.

Flascha, Carlo. "On Opium: Its History, Legacy and Cultural Benefits," *Prospect Journal*, May 25, 2011.

Frakt, Austin. "Dealing With Opioid Abuse Would Pay for Itself," *New York Times*, August 4, 2014.

Frankenburg, Frances R. *Brain-Robbers: How Alcohol, Cocaine, Nicotine, and Opiates Have Changed Human History*. Santa Barbara, California: Praeger, 2014.

Frazier, I. "The Antidote," *The New Yorker*, September 8, 2014.

Grattan, A., M. D. Sullivan, K. W. Saunders, et al. "Depression and Prescription Opioid Misuse Among Chronic Opioid Therapy Recipients with No History of Substance Abuse," *Annals of Family Medicine* 10 (2012): 304–311.

Harris, Nancy. *Opiates*. Farmington Hills, Michigan: Greenhaven Press, 2005.

Jayawant, S. S., and R. Balkrishnana. "The Controversy Surrounding OxyContin Abuse: Issues and Solutions," *Therapeutics and Clinical Risk Management* 1 (2005): 77–82.

Katz, D. A., and L. R. Hays. "Adolescent OxyContin Abuse," *Journal of the Academy of Adolescent Psychiatry* 43 (2004): 231–234.

Kolata, G., and S. Cohen. "Drug Overdoses Propel Rise in Mortality Rates in Young Whites," *New York Times*, January 16, 2016.

Meier, B. "Overdoses of Painkiller Are Linked to 282 Deaths," *New York Times*, October 28, 2001.

Meier, B. "At Painkiller Trouble Spot, Signs Seen as Alarming Didn't Alarm Drug's Maker," *New York Times*, December 10, 2001.

Meier, B. "Official Faults Drug Company for Marketing of Its Painkiller," *New York Times,* December 12, 2001.

Meier, B. "Doctor to Face U.S. Charges in Drug Case," *New York Times,* December 23, 2001.

Meier, B. "OxyContin Prescribers Face Charges in Fatal Overdoses," *New York Times,* January 19, 2002.

Meier, B. "A Small-Town Clinic Looms Large as a Top Source of Disputed Painkillers," *New York Times,* February 10, 2002.

Meier, B. "OxyContin Deaths May Top Early Count," *New York Times,* April 15, 2002.

Meier, Barry. *Pain Killer: A "Wonder" Drug's Trail of Addiction and Death.* New York: Vook, 2013.

Meier, Barry. *A World Full of Hurt: Fixing Pain Medicine's Biggest Mistake.* New York: The New York Times Company, 2013.

Meldrum, Marcia L. *Opioids and Pain Relief: A Historical Perspective.* Seattle: IASP Press, 2003.

Mundell, E. J. "FDA OK's 'Abuse Deterrent' Label for New Oxycontin," *HealthDay News,* April 16, 2010.

Nicolaou, K. C., and T. Montagnon. *Molecules That Changed the World: A Brief History of the Art and Science of Synthesis and Its Impact on Society.* Weinheim, Germany: Wiley-VCH Verlag GmbH & Co., 2008.

Paulozzi, L., G. Baldwin, G. Franklin, et al. "CDC Grand Rounds: Prescription Drug Overdoses—a U.S. Epidemic," *Morbidity and Mortality Weekly Report* 61 (2012): 10–13.

Perrone, M. "U.S. Struggles to Limit Painkillers," *Philadelphia Inquirer,* December 20, 2015.

Poitras, G. "Oxycontin, Prescription Opioid Abuse and Economic Medicalization," *Medicolegal and Bioethics* 2 (2012): 31–43.

Portenoy, R. K., and K. M. Foley. "Chronic Use of Opioid Analgesics in Non-Malignant Pain: Report of 38 Cases," *Pain* 25 (1986): 171–186.

Quinones, Sam. *Dreamland: The True Tale of America's Opiate Epidemic.* New York: Bloomsbury Press, 2015.

Rosenblatt, R. A., and M. Caitlin. "Opioids for Chronic Pain: First Do No Harm," *Annals of Family Medicine* 10 (2012): 300–301.

Rudd, R. A., N. Aleshire, J. E. Zibell, and R. M. Gladden. "Increases in Drug and Opioid Overdose Deaths—United States, 2000–2014," *Morbidity and Mortality Weekly Report* 64 (2016): 1378–1382.

Tavernese, Sabrina. "C.D.C. Painkiller Guidelines Aim to Reduce Addiction Risk," *New York Times,* March 15, 2016.

van Zee, A. "The Promotion and Marketing of OxyContin: Commercial Triumph, Public Health Tragedy," *American Journal of Public Health* 99 (2009): 221–227.

Volkow, N. D., and A. T. McLellan. "Opioid Abuse in Chronic Pain—Misconceptions and Mitigation Strategies," *New England Journal of Medicine* 374 (2016): 1253–1263.

Wikipedia. "Felix Hoffmann," https://en.wikipedia.org/wiki/Felix_Hoffmann.

Zack, I. "Pain in the Asset," *Forbes,* February 5, 2001.

Zweifler, J. A. "Objective Evidence of Severe Disease: Opioid Use in Chronic Pain," *Annals of Family Medicine* 10 (2012): 366–368.

## THE GREAT MARGARINE MISTAKE

Aro, A., A. F. M. Kardinaal, I. Salminen, et al. "Adipose Tissue, Isomeric Trans Fatty Acids, and Risk of Myocardial Infarction in Nine Countries: The EURAMIC Study," *Lancet* 345 (1995): 273–278.

Ascherio, A., C. H. Hennekens, J. E. Buring, et al. "*Trans*-Fatty Acids Intake and Risk of Myocardial Infarction," *Circulation* 89 (1994): 94–101.

Ascherio, A., E. B. Rimm, E. L. Giovannucci, et al. "Dietary Fat and Risk of Coronary Heart Disease in Men: Cohort and Follow Up Study in the United States," *British Medical Journal* 313 (1996): 84–90.

Baylin, A., E. K. Kabagambe, A. Ascherio, et al. "High 18:2 Trans-Fatty Acids in Adipose Tissue Are Associated with Increased Risk of Nonfatal Acute Myocardial Infarction in Costa Rican Adults," *Journal of Nutrition* 133 (2003): 1186–1191.

Clifton, P. M., J. B. Keogh, and M. Noakes. "Trans Fatty Acids in Adipose Tissue and the Food Supply Are Associated with Myocardial Infarction," *Journal of Nutrition* 134 (2004): 874–879.

Cowley, R., D. Gibson, and C. Sewell, "History of Eating in the United States: Margarine Vs. Butter," http://historyofeating.umwblogs.org/butter.

Dijkstra, A. J., R. J. Hamilton, and W. Hamm (eds.). *Trans Fatty Acids*. Oxford: Blackwell Publishing, 2008.

Downs, S. M., A. M. Thow, and S. R. Leeder. "The Effectiveness of Policies for Reducing Dietary Trans Fat: A Systematic Review of the Evidence," *Bulletin of the World Health Organization* 91 (2013): 262–269.

Ferdman, R. A. "The Generational Battle of Butter Vs. Margarine," *Washington Post*, June 17, 2014.

Gorelick, R. "FDA Trans-Fat Ban Threatens Berger Cookies," *Baltimore Sun*, November 22, 2013.

Hallock, B. "Rise and Fall of Trans Fat: A History of Partially Hydrogenated Oil," *Los Angeles Times*, November 7, 2013.

Harvard University School of Public Health, "Shining the Spotlight on Trans Fats," http://hsph.harvard.edu/nutritionsource/transfats/#big_changes.

Hu, F. B., M. J. Stampfer, J. E. Manson, et al. "Dietary Fat Intake and the Risk of Coronary Heart Disease in Women," *New England Journal of Medicine* 337 (1997): 1491–1499.

Katan, M. B., P. L. Zock, and R. P. Mensink. "Trans Fatty Acids and Their Effects on Lipoproteins in Humans," *Annual Reviews of Nutrition* 15 (1995): 473–493.

Khazan, O. "When Trans Fats Were Healthy," *The Atlantic,* November 8, 2013.

Kolata, G. "Mediterranean Diet Shown to Ward Off Heart Attack and Stroke," *New York Times,* February 25, 2013.

Lemaitre, R. N., I. B. King, T. E. Raghunathan, et al. "Cell Membrane *Trans*-Fatty Acids and the Risk of Primary Cardiac Arrest," *Circulation* 105 (2002): 697–701.

Mensink, R. P., P. L. Zock, A. D. M. Kester, and M. B. Katan. "Effects of Dietary Fatty Acids and Carbohydrates on the Ratio of Serum Total to HDL Cholesterol and on Serum Lipids and Apolipoproteins: A Meta-Analysis of 60 Controlled Studies," *American Journal of Clinical Nutrition* 77 (2003): 1146–1155.

Mozaffarian, D., M. B. Katan, A. Ascherio, et al. "Trans Fatty Acids and Cardiovascular Disease," *New England Journal of Medicine* 354 (2006): 1601–1613.

O'Connor, A. "Study Questions Fat and Heart Disease Link," *New York Times,* March 17, 2014.

Oh, D., F. B. Hu, J. E. Manson, et al. "Dietary Fat Intake and Risk of Coronary Heart Disease in Women: 20 Years of Follow-Up of the Nurses Health Study," *American Journal of Epidemiology* 161 (2005): 672–679.

Oomen, C. M., M. C. Ocké, E. J. M. Feskens, et al. "Association Between Trans Fatty Acid Intake and 10-Year Risk of Coronary Heart Disease in the Zutphen Elderly Study: A Prospective, Population-Based Study," *Lancet* 357 (2001): 746–751.

Pietinen, P., A. Ascherio, P. Korhonen, et al. "Intake of Fatty Acids and Risk of Coronary Heart Disease in a Cohort of Finnish Men: the Alpha-Tocopherol, Beta-Carotene Cancer Prevention Study," *American Journal of Epidemiology* 145 (1997): 876–887.

Remig, V., B. Franklin, S. Margolis, et al. "Trans Fats in America: A Review of Their Use, Consumption, Health Implications, and Regulation," *Journal of the American Dietetic Association* 110 (2010): 585–592.

Ross, J. K. "The FDA Wants to Ban Berger Cookies, the World's Most Delicious Dessert," www.reason.com, November 23, 2013.

Rothstein, William G. *Public Health and the Risk Factor: A History of Uneven Medical Revolution.* Rochester, New York: University of Rochester Press, 2003.

Schleifer, D. "Fear of Frying: A Brief History of Trans Fats," https://nplusonemag.com/online-only/online-only/fear-frying/.

Schleifer, D. "The Perfect Solution: How Trans Fats Became the Healthy Replacement for Saturated Fats," *Technology and Culture* 53 (2012): 94–119.

Shaw, Judith. *Trans Fats: The Hidden Killer in Our Food.* New York: Pocket Books, 2004.

Stampfer, M. J., F. M. Sacks, S. Salvini, et al. "A Prospective Study of Cholesterol, Apolipoproteins, and the Risk of Myocardial Infarction," *New England Journal of Medicine* 325 (1991): 373–381.

Stender, S., and J. Dyerberg. "High Levels of Industrially Produced Trans Fats in Popular Fast Foods," *New England Journal of Medicine* 354 (2006): 1650–1652.

Stender, S., A. Astrup, and J. Dyerberg. "Ruminant and Industrially Produced Trans Fatty Acids: Health Aspects," *Food & Nutrition Research*, March 12, 2008, doi:10.3402/fnr.v52i0.1651.

Taubes, G. "The Soft Science of Dietary Fat," *Science* 291 (2001): 2536–2545.

Thomas, L. H., P. R. Jones, J. A. Winter, and H. Smith. "Hydrogenated Oils and Fats: The Presence of Chemically-Modified Fatty Acids in Human Adipose Tissue," *American Journal of Clinical Nutrition* 34 (1981): 877–886.

United States Food and Drug Administration. "Trans Fat Now Listed with Saturated Fat and Cholesterol," www.fda.gov/Food/Ingredients PackagingLabelling/Nutrition/ucm274590.htm.

van Tol, A., P. L. Zock, T. van Gent, et al. "Dietary *Trans* Fatty Acids Increase Serum Cholesterylester Transfer Protein Activity in Man," *Atherosclerosis* 115 (1995): 129–134.

Whoriskey, P. "The U.S. Government Is Poised to Withdraw Longstanding Warnings About Cholesterol," *Washington Post,* February 10, 2015.

Willett, W. C., M. J. Stampfer, J. E. Manson, et al. "Intake of Trans Fatty Acids and Risk of Coronary Heart Disease Among Women," *Lancet* 341 (1993): 581–585.

## BLOOD FROM AIR

Charles, Daniel. *Between Genius and Genocide: The Tragedy of Fritz Haber, Father of Chemical Warfare*. London: Jonathan Cape, 2005.

Goran, Morris. *The Story of Fritz Haber.* Norman: University of Oklahoma Press, 1967.

Haber, L. F. *The Poisonous Cloud: Chemical Warfare in the First World War*. Oxford: Clarendon Press, 1986.

Hager, Thomas. *The Alchemy of Air: A Jewish Genius, a Doomed Tycoon, and the Scientific Discovery that Fed the World and Fueled the Rise of Hitler*. New York: Broadway Books, 2008.

Smil, Vaclav. *Enriching the Earth: Fritz Haber, Carl Bosch, and the Transformation of World Food Production.* Cambridge: MIT Press, 2001.

Stern, Fritz. *Dreams and Delusions: The Drama of German History.* New Haven: Yale University Press, 1999.

Stoltzenberg, Dietrich. *Fritz Haber: Chemist, Nobel Laureate, German, Jew.* Philadelphia: Chemical Heritage Press, 2004.

## AMERICA'S MASTER RACE

### The Republican Primary

Clement, Scott. "Republicans Embrace Trump's Ban on Muslims While Most Others Reject It," *Washington Post,* December 14, 2015.

Cruz, Ted. "Cruz Immigration Plan," www.tedcruz.org/cruz-immigration-plan.

Haberman, Maggie. "Donald Trump Deflects Withering Fire on Muslim Plan," *New York Times,* December 8, 2015.

Hussain, Murtaza. "Majority of Americans Now Support Donald Trump's Proposed Muslim Ban, Poll Shows," *The Intercept,* March 30, 2016.

Osnos, Evan. "The Fearful and the Frustrated," *The New Yorker,* August 31, 2015.

Savage, Charlie. "Plan to Bar Foreign Muslims by Donald Trump Might Survive a Lawsuit," *New York Times,* December 8, 2015.

Ye Hee Lee, Michelle. "Donald Trump's False Comments Connecting Mexican Immigrants and Crime," *Washington Post,* July 8, 2015.

**Madison Grant and *The Passing of the Great Race***

Bryson, Bill. *One Summer: America, 1927*. New York: Doubleday, 2013.

Black, Edwin. *War Against the Weak: Eugenics and America's Campaign to Create a Master Race*. Washington, D.C.: Dialogue Press, 2003.

Chesterton, G. K. *Eugenics and Other Evils: An Argument Against the Scientifically Organized State*. Seattle: Inkling Books, 2000.

Cohen, Adam. *Imbeciles: The Supreme Court, American Eugenics, and the Sterilization of Carrie Buck*. New York: Penguin Press, 2016.

Grant, Madison. *The Passing of the Great Race, Or, the Racial Basis of European History*. New York: Charles Scribner's Sons, 1916.

Lagnado, Lucette Matalon, and Sheila Cohn Dekel. *Children of the Flames: Dr. Josef Mengele and the Untold Story of the Twins of Auschwitz*. New York: William Morrow and Company, 1991.

Lombardo, Paul A. *Three Generations, No Imbeciles: Eugenics, the Supreme Court, and* Buck v. Bell. Baltimore: Johns Hopkins University Press, 2008.

Lombardo, Paul A. *A Century of Eugenics in America: From the Indiana Experiment to the Human Genome Project*. Bloomington: Indiana University Press, 2011.

Naftali, Timothy. "Unlike Ike," *New York Times Book Review*, September 26, 2015.

Nourse, Victoria. "When Eugenics Became Law," *Nature* 530 (2016): 418.

Oshinsky, David. "No Justice for the Weak," *New York Times Book Review*, March 20, 2016.

Perl, Gisella. *I Was a Doctor in Auschwitz*. Tamarac, Florida: Yale Garber, 1987.

Posner, Gerald L., and John Ware. *Mengele: The Complete Story*. New York: Cooper Square Press, 1986.

Spiro, Jonathan Peter. *Defending the Master Race: Conservation, Eugenics, and the Legacy of Madison Grant*. Burlington: University of Vermont Press, 2009.

## TURNING THE MIND INSIDE OUT

Braslow, Joel. *Mental Ills and Bodily Cures: Psychiatric Treatment in the First Half of the Twentieth Century*. Berkeley: University of California Press, 1997.

Connett, David. "Autism: Potentially Lethal Bleach 'Cure' Feared to Have Spread to Britain," *The Independent*, November 23, 2015.

Dully, Howard, and Charles Fleming. *My Lobotomy*. New York: Three Rivers Press, 2007.

El-Hai, Jack. *The Lobotomist: A Maverick Medical Genius and His Tragic Quest to Rid the World of Mental Illness*. Hoboken: John Wiley & Sons, 2005.

Freeman, W., and J. W. Watts. "Prefrontal Lobotomy: The Surgical Relief of Mental Pain," *Bulletin of the New York Academy of Medicine* 18 (1942): 794–812.

Fuster, Joaquin M. *The Prefrontal Cortex: Fourth Edition*. San Diego: Academic Press, 2008.

Grimes, D. R. "Autism: How Unorthodox Treatments Can Exploit the Vulnerable," *The Guardian*, July 15, 2015.

Johnson, J. *American Lobotomy: A Rhetorical History*. Ann Arbor: University of Michigan Press, 2014.

Kent, Deborah. *Snake Pits, Talking Cures, & Magic Bullets: A History of Mental Illness*. Brookfield, Connecticut: Twenty-First Century Books, 2003.

Larson, Kate Clifford. *Rosemary: The Hidden Kennedy Daughter*. New York: Houghton Mifflin Harcourt, 2015.

Lynn, G., and E. Davey. "'Miracle Autism Cure' Seller Exposed by BBC Investigation," *BBC News*, London, June 11, 2015.

Miller, Bruce L., and Jeffrey L. Cummings (eds.). *The Human Frontal Lobes: Functions and Disorders: Second Edition*. New York: The Guilford Press, 2007.

Miller, M. E. "The Mysterious Death of a Doctor Who Peddled Autism 'Cures' to Thousands," *Washington Post,* July 16, 2015.

Momsense (blog). "The Miracle Mineral Solution Sham and What You Can Do About It," www.itsmomsense.com/mms-sham.

Nasaw, David. *The Patriarch: The Remarkable Life and Turbulent Times of Joseph P. Kennedy*. New York: Penguin Press, 2012.

Newitz, A., "The Strange, Sad History of the Lobotomy," http://io9.com/5787430/the-strange-sad-history-of-the-lobotomy.

Partridge, Maurice. *Pre-Frontal Leucotomy*. Oxford: Blackwell Scientific Publications, 1950.

Pressman, Jack D. *Last Resort: Psychosurgery and the Limits of Medicine.* Cambridge: Cambridge University Press, 1998.

Raz, Mical. *The Lobotomy Letters: The Making of American Psychosurgery*. Rochester: The University of Rochester Press, 2013.

Shorter, Edward. *A History of Psychiatry: From the Era of the Asylum to the Age of Prozac*. New York: John Wiley & Sons, 1997.

Shutts, David. *Lobotomy: Resort to the Knife*. New York: Van Norstrand Reinhold, 1982.

Valenstein, Elliot S. *Great and Desperate Cures: The Rise and Decline of Psychosurgery and Other Radical Treatments for Mental Illness.* CreateSpace Independent Publishing Platform, 2010.

Whitaker, R. *Mad in America: Bad Science, Bad Medicine, and the Enduring Mistreatment of the Mentally Ill.* New York: Basic Books, 2002.

Willingham, E. "Here's Why Authorities Searched the Offices of Controversial Autism Doctor Bradstreet," *Forbes,* July 9, 2015.

## THE MOSQUITO LIBERATION FRONT

Allen, Arthur. *The Fantastic Laboratory of Dr. Weigl: How Two Brave Scientists Battled Typhus and Sabotaged the Nazis*. New York: W. W. Norton, 2014.

Carson, Rachel. *The Sea Around Us.* New York: Oxford University Press, 1950.

Carson, Rachel. *The Edge of the Sea.* New York: Houghton Mifflin, 1955.

Carson, Rachel. *Silent Spring*. New York: Houghton Mifflin, 1962.

Carson, Rachel. *The Sense of Wonder*. New York: Harper & Row Publishers, 1965.

Darrell, Ed. "Setting the Record Straight on Rachel Carson, Malaria, and DDT," Millard Fillmore's Bathtub (blog), https://timpanogos.wordpress.com/2007/06/19/setting-the-record-straight-on-rachel-carson-malaria-and-ddt, June 19, 2007.

Driessen, Paul. *Eco-Imperialism: Green Power, Black Death*. Bellevue, Washington: Free Enterprise Press, 2003.

Dunlap, Thomas R. *DDT, Silent Spring, and the Rise of Environmentalism: Classic Texts*. Seattle: University of Washington Press, 2008.

Kinkela, David. *DDT & the American Century: Global Health, Environmental Politics, and the Pesticide that Changed the World*. Chapel Hill: The University of North Carolina Press, 2011.

Kudlinski, Kathleen V. *Rachel Carson: Pioneer of Ecology*. New York: Puffin Books, 1988.

Lawlor, Laurie. *Rachel Carson and Her Book That Changed the World.* New York: Holiday House, 2012.

Lear, Linda. *Rachel Carson: Witness for Nature*. New York: Houghton Mifflin Harcourt, 1997.

Lear, Linda. *Lost Woods: The Discovered Writing of Rachel Carson*. Boston: Beacon Press, 1998.

Lytle, Mark Hamilton. *The Gentle Subversive: Rachel Carson, Silent Spring, and the Rise of the Environmental Movement*. New York: Oxford University Press, 2007.

Meiners, Roger, Pierre Desrochers, and Andrew Morriss (eds.). *Silent Spring at 50: The False Crisis of Rachel Carson*. Washington, D.C.: Cato Institute, 2012.

Musil, Robert K. *Rachel Carson and Her Sisters: Extraordinary Women Who Have Shaped America's Environment*. New Brunswick, New Jersey: Rutgers University Press, 2014.

Oreskes, Naomi, and Erik K. Conway. *Merchants of Doubt: How a Handful of Scientists Obscured the Truth on Issues from Tobacco Smoke to Global Warming*. New York: Bloomsbury Press, 2010.

Pearson, Gwen. "DDT, Junk Science, Malaria, and Insecticide Resistance," https://membracid.wordpress.com/2007/06/13/ddt-malaria-insecticide-resistance, June 13, 2007.

Pearson, Gwen. "Setting the Record Straight on Rachel Carson," https://membracid.wordpress.com/2007/06/25/setting-the-record-straight-on-rachel-carson, June 25, 2007.

Roberts, Donald, Richard Tren, Roger Bate, and Jennifer Zambone. *The Excellent Powder: DDT's Political and Scientific History*. Indianapolis: Dog Ear Publishing, 2010.

Souder, William. *On a Farther Shore: The Life and Legacy of Rachel Carson*. New York: Broadway Books, 2012.

Strickman, Daniel, Stephen P. Frances, and Mustapha Debboun. *Prevention of Bug Bites, Stings, and Disease*. New York: Oxford University Press, 2009.

## NOBEL PRIZE DISEASE

### Linus Pauling

Goertzel, Ted, and Ben Goertzel. *Linus Pauling: A Life in Science and Politics*. New York: Basic Books, 1995.

Hager, Thomas. *Force of Nature: The Life of Linus Pauling*. New York: Simon & Schuster, 1995.

Hager, Thomas. *Linus Pauling and the Chemistry of Life*. Oxford: Oxford University Press, 1998.

Marinacci, Barbara (ed.). *Linus Pauling in His Own Words: Selections from His Writings, Speeches, and Interviews*. New York: Simon & Schuster, 1995.

Mead, Clifford, and Thomas Hager (eds.). *Linus Pauling: Scientist and Peacemaker*. Corvallis: Oregon State University Press, 2001.

Newton, David E. *Linus Pauling: Scientist and Advocate*. New York: Facts on File, 1994.

Offit, Paul. *Do You Believe in Magic? The Sense and Nonsense of Alternative Medicine*. New York: HarperCollins, 2013.

Pauling, Linus. *Vitamin C and the Common Cold*. San Francisco: W. H. Freeman and Company, 1970.

Pauling, Linus. *Vitamin C, the Common Cold, and the Flu*. San Francisco: W. H. Freeman and Company, 1976.

Pauling, Linus, and Ewan Cameron (eds.). *Cancer and Vitamin C: A Discussion of the Nature, Causes, Prevention, and Treatment of Cancer with Special Reference to the Value of Vitamin C*. Philadelphia: Camino Books, 1979 (updated 1993).

Pauling, Linus. *How to Live Longer and Feel Better*. Corvallis: Oregon State University Press, 1986.

Price, Catherine. *Vitamania: Our Obsessive Quest for Nutritional Perfection*. New York: Penguin Press, 2015.

Serafini, Anthony. *Linus Pauling: A Man and His Science*. Lincoln,

Nebraska: Paragon House Publishers, 1989.

Sherrow, Victoria. *Linus Pauling: Investigating the Magic Within*. Austin: Raintree Steck-Vaughn Publishers, 1997.

Valiunas, A. "The Man Who Thought of Everything," *The New Atlantis*, Spring 2015.

**Peter Duesberg**

Cohen, J. "The Duesberg Phenomenon," *Science* 266 (1994): 1642–1644.

Cohen, J. "Duesberg and Critics Agree: Hemophilia Is the Best Test," *Science* 266 (1994): 1645–1646.

Cohen, J. "Fulfilling Koch's Postulates," *Science* 266 (1994): 1647.

Cohen, J. "Could Drugs, Rather Than a Virus, Be the Cause of AIDS?" *Science* 266 (1994): 1648–1649.

Kalichman, Seth. *Denying AIDS: Conspiracy Theories, Pseudoscience, and Human Tragedy*. New York: Copernicus Books, 2009.

Nattrass, Nicoli. *The AIDS Conspiracy: Science Fights Back*. New York: Columbia University Press, 2012.

**Luc Montagnier**

Butler, D. "Trial Draws Fire," *Nature* 468 (2010): 743.

Enserink, M. "French Nobelist Escapes 'Intellectual Terror' to Pursue Radical Ideas in China," *Science* 330 (2010): 1732.

Gorski, D. "Luc Montagnier: The Nobel Disease Strikes Again," http://scienceblogs.com/insolence/2010/11/23/luc-montagnier-the-nobel-disease-strikes, November 23, 2010.

Gorski, D. "The Nobel Disease Meets DNA Teleportation and Homeopathy," http://scienceblogs.com/?s=the+nobel+prize+meets+dna+teleportation, January 14, 2011.

Gorski, D. "Luc Montagnier and the Nobel Disease," June 4, 2012, http://www.sciencebasedmedicine.org/luc-montagnier-and-the-nobel-disease.

Montagnier, L. "Autism: The Microbial Track," www.autismone.org/content/keynote-microbial-track.

Salzberg, S. "Nobel Laureate Joins Anti-Vaccination Crowd at Autism One," *Forbes,* May 27, 2012.

Ullman, D. "Luc Montagnier, Nobel Prize Winner, Takes Homeopathy Seriously," *Huffington Post,* January 30, 2011.

## LEARNING FROM THE PAST

### MMR Vaccine and Autism

Chen, R.T., and F. DeStefano. "Vaccine Adverse Events: Causal or Coincidental?" *Lancet* 351 (1998): 611–612.

Dales, L., S. J. Hammer, and N. J. Smith, "Time Trends in Autism and in MMR Immunization Coverage in California," *Journal of the American Medical Association* 285 (2001): 1183–1185.

Davis, R. L., P. Kramarz, K. Bohlke, et al. "Measles-Mumps-Rubella and Other Measles-Containing Vaccines Do Not Increase the Risk for Inflammatory Bowel Disease: a Case-Control Study from the Vaccine Safety Datalink Project," *Archives of Pediatric Adolescent Medicine* 155 (2001): 354–359.

DeStefano, F., and W. W. Thompson. "MMR Vaccine and Autism: an Update of the Scientific Evidence," *Expert Review of Vaccines* 3 (2004): 19–22.

DeStefano, F., T. K. Bhasin, W. W. Thompson, et al. "Age at First Measles-Mumps-Rubella Vaccination in Children with Autism and School-Matched Control Subjects: a Population-Based Study in Metropolitan Atlanta," *Pediatrics* 113 (2004): 259–266.

Farrington, C. P., E. Miller, and B. Taylor. "MMR and Autism: Further Evidence Against a Causal Association," *Vaccine* 19 (2001): 3632–3635.

Fombonne, E., and S. Chakrabarti. "No Evidence for a New Variant of Measles-Mumps-Rubella-Induced Autism," *Pediatrics* 108 (2001): E58.

Honda, H., Y. Shimizu, and M. Rutter, "No Effect of MMR Withdrawal on the Incidence of Autism: a Total Population Study," *Journal of Child Psychology and Psychiatry* 4 (2005): 572–579.

Kaye, J. A., M. Mar Melero-Montes, and H. Jick, "Mumps, Measles, and Rubella Vaccine and the Incidence of Autism Recorded by General Practitioners: a Time Trend Analysis," *British Medical Journal* 322 (2001): 460–463.

Madsen, K. M., and M. Vestergaard. "MMR Vaccination and Autism: What Is the Evidence for a Causal Association?" *Drug Safety* 27 (2004): 831–840.

Miller, E. "Measles-Mumps-Rubella Vaccine and the Development of Autism," *Seminars in Pediatric Infectious Diseases* 14 (2003): 199–206.

Public Health Laboratory Service. "Measles Outbreak in London," *Communicable Diseases Report Weekly* 12 (2002): 1.

Stratton, K., A. Gable, and P. M. M. Shetty. "Measles-Mumps-Rubella Vaccine and Autism," Institute of Medicine, Immunization Safety Review Committee. Washington, D.C.: National Academies Press, 2001.

Taylor, B., E. Miller, C. P. Farrington, et al. "Autism and Measles, Mumps, and Rubella Vaccine: No Epidemiological Evidence for a Causal Association," *Lancet* 353 (1999): 2026–2029.

Taylor, B., E. Miller, R. Lingam, et al. "Measles, Mumps, and Rubella Vaccination and Bowel Problems or Developmental Regression in

Children with Autism: a Population Study," *British Medical Journal* 324 (2002): 393–396.

Wakefield, A. J., S. H. Murch, A. Anthony, et al. "Ileal-Lymphoid-Nodular Hyperplasia, Non-Specific Colitis, and Pervasive Developmental Disorder in Children," *Lancet* 351 (1998): 637–641 (retracted).

Weiss, S. "Eat Dirt—The Hygiene Hypothesis and Allergic Diseases," *New England Journal of Medicine* 347 (2002): 390–391.

Wilson K., E. Mills, C. Ross, et al. "Association of Autistic Spectrum Disorder and the Measles, Mumps, and Rubella Vaccine: a Systematic Review of Current Epidemiological Evidence," *Archives of Pediatric and Adolescent Medicine* 157 (2003): 628–634.

**Thimerosal**

Andrews, N., E. Miller, A. Grant, et al. "Thimerosal Exposure in Infants and Developmental Disorders: a Retrospective Cohort Study in the United Kingdom Does Not Show a Causal Association," *Pediatrics* 114 (2004): 584–591.

Centers for Disease Control and Prevention. "Thimerosal in Vaccines: a Joint Statement of the American Academy of Pediatrics and the Public Health Service," *Morbidity and Mortality Weekly Report* 48 (1999): 563–565.

Centers for Disease Control and Prevention. "Recommendations Regarding the Use of Vaccines that Contain Thimerosal as a Preservative," *Morbidity and Mortality Weekly Report* 48 (1999): 996–998.

Clark, S. J., M. D. Cabana, T. Malik, et al. "Hepatitis B Vaccination Practices in Hospital Newborn Nurseries Before and After Changes in Vaccination Recommendations," *Archives of Pediatric Adolescent Medicine* 155 (2001): 915–920.

Fombonne, E., R. Zakarian, A. Bennett, et al. "Pervasive Developmental Disorders in Montreal, Quebec, Canada: Prevalence and Links with Immunization," *Pediatrics* 118 (2006): 139–150.

Gundacker, C., B. Pietschnig, K. J. Wittmann, et al. "Lead and Mercury in Breast Milk," *Pediatrics* 110 (2002): 873–878.

Heron, J., J. Golding, and the ALSPAC Study Team. "Thimerosal Exposure in Infants and Developmental Disorders: a Prospective Cohort Study in the United Kingdom Does Not Show a Causal Association," *Pediatrics* 114 (2004): 577–583.

Hviid, A., M. Stellfeld, J. Wohlfahrt, et al. "Association Between Thimerosal-Containing Vaccine and Autism," *Journal of the American Medical Association* 290 (2003): 1763–1766.

Institute of Medicine (US) Immunization Safety Review Committee, D. Stratton, A. Gable, and M. C. McCormick (eds.). *Immunization Safety Review: Thimerosal-Containing Vaccines and Neurodevelopmental Disorders.* Washington, D.C.: National Academies Press, 2001.

Marsh, D. O., T. W. Clarkson, C. Cox, et al. "Fetal Methylmercury Poisoning: Relationship Between Concentration in Single Strands of Maternal Hair and Child Effects," *Archives of Neurology* 44 (1987): 1017–1022.

Nelson, K. B., and M. L. Bauman. "Thimerosal and Autism?" *Pediatrics* 111 (2003): 664–679.

Parker, S. K., B. Schwartz, J. Todd, et al. "Thimerosal-Containing Vaccines and Autistic Spectrum Disorder: a Critical Review of Published Original Data," *Pediatrics* 114 (2004): 793–804.

Thompson, W. W., C. Price, B. Goodson, et al. "Early Thimerosal Exposure and Neuropsychological Outcomes at 7 to 10 Years," *New England Journal of Medicine* 357 (2007): 1281–1292.

Verstraeten, T., R. L. Davis, F. DeStefano, et al. "Safety of Thimerosal-Containing Vaccines: a Two-Phased Study of Computerized Health Maintenance Organization Databases," *Pediatrics* 112 (2003): 1039–1048.

## E-cigarettes

American Academy of Pediatrics. "Electronic Nicotine Delivery Systems," *Pediatrics* 136 (2015): 1018–1026.

Brown, J., E. Beard, D. Kotz, et al. "Real-World Effectiveness of E-Cigarettes When Used to Aid Smoking Cessation: A Cross-Sectional Population Study," *Addiction* 109 (2014): 1531–1540.

Clarke, T. "Youth E-Cigarette Data Prompts Call to Speed Regulation," Reuters, April 18, 2015.

Davidson, L. "Vaping Takes Off as E-Cigarettes Break Through $6BN," *Telegraph,* June 23, 2015.

Farsalinos, K. E., and R. Poisa. "Safety Evaluation and Risk Assessment of Electronic Cigarettes as Tobacco Cigarette Substitutes: A Systematic Review," *Therapeutic Advances in Drug Safety* 5 (2014): 67–86.

Friedman, A. S. "How Does Electronic Cigarette Access Affect Adolescent Smoking?" *Journal of Health Economics,* October 19, 2015, doi:10.1016/j.healeco.2015.10.003.

Green, S. H., R. Bayer, and A. L. Fairchild. "Evidence, Policy, and E-Cigarettes—Will England Reframe the Debate," *New England Journal of Medicine* 374 (2016): 1301–1303.

Haelle, T. "Teen Vaping Triples: E-Cigarettes, Hookahs Threaten Drop in Teen Tobacco Use," *Forbes,* April 17, 2015.

Haelle, T. "E-Cigarettes Benefit Public Health If Used to Replace Smoking, Say British Doctors," *Forbes,* April 28, 2016.

Herzog, B. "Will Electronic Cigarettes Pass Combustible Cigarette

Sales Within the Next 10 Years," www.breatheic.com/blog/will-electronic-cigarette-pass-combustible-cigarette-sales-within-the-next-10-years-2/.

Jamal, A., I. T. Israel, E. O'Connor, et al. "Current Cigarette Smoking Among Adults—United States, 2005–2013," *Morbidity and Mortality Weekly Report* 63 (2014): 1108–1112.

Jamal, A. J., D. M. Homa, E. O'Connor, et al. "Current Cigarette Smoking Among Adults—United States, 2005–2014," *Morbidity and Mortality Weekly Report* 64 (2015): 1235–1240.

Klein, J. D. "Electronic Cigarettes Are Another Route to Nicotine Addiction for Youth," *Journal of the American Medical Association Pediatrics*, September 8, 2015, doi:10.1001/jamapediatrics.2015.1929.

McNeill, A. B., C. R. Hitchman, P. Hajek, and H. McRobbie. *E-Cigarettes: An Evidence Update, A Report Commissioned by Public Health England.* Public Health England, August 2015.

Nitzkin, J. "E-Cigarettes: A Life Saving Technology or a Way for Tobacco Companies to Re-Normalize Smoking in American Society," *The Food and Drug Law Institute* 4 (2014): 1–17.

Nitzkin, J. "Understanding the Crusade Against E-Cigarettes," rstreet.org, November 23, 2015.

Nocera, J. "Is Vaping Worse Than Smoking?" *New York Times,* January 27, 2015.

Satel, S. "What's Driving the War on E-Cigarettes?" *National Review,* June 1, 2015.

Satel, S. "The Year in E-Cigarettes: The Good, the Bad, the Reason for Optimism," *Forbes,* December 31, 2015.

**Bisphenol A**

Groopman, J. "The Plastic Panic: How Worried Should We Be About Everyday Chemicals," *The New Yorker,* May 31, 2010.

Hall, H. "Phthalates and BPA: Of Mice and Men," Science-Based Medicine, December 13, 2011.

Hengstler, J. G., H. Foth, T. Gebel, et al. "Critical Evaluation of Key Evidence on the Human Health Hazards of Exposure to Bisphenol A," *Critical Reviews in Toxicology* 41 (2011): 263–291.

Hinterthuer, A. "Just How Harmful Are Bisphenol A Plastics?" *Scientific American,* September 2008.

Kennedy, L. "Bisphenol A Is Harmless," The Skeptics Society Forum, March 11, 2013.

**Cancer Screening**

Ablin, R. J. "The Great Prostate Mistake," *New York Times,* March 9, 2010.

Ahn, H. S., H. J. Kim, and H. G. Welch. "Korea's Thyroid Cancer 'Epidemic'—Screening and Overdiagnosis," *New England Journal of Medicine* 371 (2014): 1765–1767.

Ahn, H. S., and H. G. Welch. "South Korea's Thyroid Cancer 'Epidemic'—Turning the Tide," *New England Journal of Medicine* 373 (2015): 2389–2390.

Ashwanden, C. "Why I'm Opting Out of Mammography," *Journal of the American Medical Association Internal Medicine* 175 (2015): 164–165.

Bangma, C. H., S. Roemeling, and F. H. Schröder. "Overdiagnosis and Overtreatment of Early Detected Prostate Cancer," *World Journal of Urology* 25 (2007): 3–9.

Bernstein, Lenny. "After New Guidelines, U.S. Sees Sharp Decline in Prostate Cancer Screenings—And Diagnoses," *Washington Post,* November 17, 2015.

Bleyer, A., and H. G. Welch, "Effect of Three Decades of Screening Mammography on Breast-Cancer Incidence," *New England Journal of Medicine* 367 (2012): 1998–2005.

Elmore, J. G., and R. Etzioni. "Effect of Screening Mammography on Cancer Incidence and Mortality," *Journal of the American Medical Association Internal Medicine* 175 (2015): 1490–1491.

Esserman, L., Y. Shieh, and I. Thompson. "Rethinking Screening for Breast Cancer and Prostate Cancer," *Journal of the American Medical Association* 302 (2009): 1685–1692.

Etzioni, R., D. F. Penson, J. M. Legler, et al. "Overdiagnosis Due to Prostate-Specific Antigen Screening: Lessons from U.S. Prostate Cancer Incidence Trends," *Journal of the National Cancer Institute* 94 (2002): 981–990.

Garas, G., A. Qureishi, F. Palazzo, et al. "Should We Be Operating on All Thyroid Cancers?" Paper presented at the Fifth Congress of the International Federation of Head and Neck Oncological Societies, July 26–30, 2014, New York, Abstract P0085.

Gawande, A. "Overkill: An Avalanche of Unnecessary Medical Care Is Harming Patients Physically and Financially. What Can We Do About It?" *The New Yorker,* May 11, 2015.

Grady, Denise. "Early Prostate Cancer Cases Fall Along with Screening," *New York Times,* November 17, 2015.

Hafner, Katie. "A Breast Cancer Surgeon Who Keeps Challenging the Status Quo," *New York Times,* September 28, 2015.

Harding, C., F. Pompei, D. Burmistrov, et al. "Breast Cancer Screening, Incidence, and Mortality Across US Counties," *Journal of the American Medical Association Internal Medicine* 175 (2015): 1483–1489.

Kaplan, K. "Screening Mammograms Don't Prevent Breast Cancer Deaths," *Los Angeles Times,* July 6, 2015.

Kolata, G. "Study Points to Overdiagnosis of Thyroid Cancer," *New York Times,* November 5, 2014.

Kolata, G. "It's Not Cancer: Doctors Reclassify a Thyroid Tumor," *New York Times,* April 14, 2016.

Lee, J-H, and S. W. Shin. "Overdiagnosis and Screening of Thyroid Cancer in Korea," *Lancet* 384 (2014): 1848.

McCullough, M. "When Mammograms Are More Harm Than Help," *Philadelphia Inquirer*, July 12, 2015.

Moyer, V. A., on behalf of the U.S. Preventive Services Task Force. "Screening for Prostate Cancer: U.S. Preventive Services Task Force Recommendation Statement," *Annals of Internal Medicine* 157 (2012): 120–134.

Narod, S. A., J. Iqbal, V. Giannakeas, et al. "Breast Cancer Mortality After a Diagnosis of Ductal Carcinoma *In Situ*," *Journal of the American Medical Association Oncology*, August 20, 2015.

Penson, D. "The Pendulum of Prostate Cancer Screening," *Journal of the American Medical Association* 314 (2015): 2031–2033.

Rapaport, L. "Less Frequent Cancer Screenings Possible for Many People, Doctor Says," Reuters, May 18, 2015.

Sammon, J. D., F. Abdollah, T. K. Choueiri, et al. "Prostate-Specific Antigen Screening After 2012 US Preventive Services Task Force Recommendation," *Journal of the American Medical Association* 314 (2015): 2077–2079.

Shute, N. "Overdiagnosis Could Be Behind Jump in Thyroid Cancer Cases," National Public Radio, February 21, 2014.

Shute, N. "More Mammograms May Not Always Mean Fewer Cancer Deaths," National Public Radio, July 7, 2015.

Tanner, L. "Less Prostate Cancer and Screening After New Guidance," Associated Press, November 17, 2015.

Volmer, R. T. "Revisiting Overdiagnosis and Fatality in Thyroid Cancer," *American Journal of Clinical Pathology* 141 (2014): 128–132.

Welch, H. G., and W. C. Black. "Overdiagnosis in Cancer," *Journal of the National Cancer Institute* 102 (2009): 605–613.

Welch, H. G., and P. C. Albertson. "Prostate Cancer Diagnosis and Treatment After the Introduction of Prostate-Specific Antigen Screening: 1986–2005," *Journal of the National Cancer Institute* 101 (2009): 1325–1329.

Welch, H. G., D. H. Gorski, and P. C. Albertson. "Trends in Metastatic Breast and Prostate Cancer—Lessons in Cancer Dynamics," *New England Journal of Medicine* 373 (2015): 1685–1687.

## Genetically Modified Organisms

Ewen, S. W., and A. Pusztai. "Effect of Diets Containing Genetically Modified Potatoes Expressing *Galanthu nivalis* Lectin on Rat Small Intestine," *Lancet* 354 (1999): 1353–1354.

Flam, F. "Defying Science and Common Sense, New York Bill Would Ban GMOs in Vaccines," *Forbes,* February 26, 2015.

Klumper W., and M. Qaim. "A Meta-Analysis of the Impacts of Genetically Modified Crops," *PLOS One* 9 (2014): 1–7.

Novella, S. "No Health Risks from GMOs," *Skeptical Inquirer,* July/August, 2014.

# ABOUT THE AUTHOR

Paul A. Offit, M.D., is the director of the Vaccine Education Center at the Children's Hospital of Philadelphia and the Maurice R. Hilleman Professor of Vaccinology and a professor of pediatrics at the Perelman School of Medicine at the University of Pennsylvania. He is the recipient of many awards, including the J. Edmund Bradley Prize for Excellence in Pediatrics from the University of Maryland Medical School, the President's Certificate for Outstanding Service from the American Academy of Pediatrics, the David E. Rogers Award from the American Association of Medical Colleges, the Odyssey Award from the Center for Medicine in the Public Interest, and the Maxwell Finland Award for Scientific Achievement from the National Foundation for Infectious Diseases.

Dr. Offit has published more than 160 papers in medical and scientific journals in the areas of rotavirus-specific immune responses and vaccine safety, and he is the co-inventor of the rotavirus vaccine RotaTeq, recommended for universal use in infants by the Centers for Disease Control and Prevention; for this achievement he was honored by Bill and Melinda Gates during the launch of their foundation's Living Proof Project for global health and was elected to the Institute of Medicine of the National Academy of Sciences. In 2015 Dr. Offit was elected to the American Academy of Arts and Sciences. He is the author of six medical narratives: *The Cutter Incident: How America's First Polio Vaccine Led to Today's Growing Vaccine Crisis; Vaccinated: One Man's Quest to Defeat the World's Deadliest Diseases,* for which he won an award from the American Medical Writers Association; *Autism's False Prophets: Bad Science, Risky Medicine, and the Search for a Cure; Deadly Choices: How the Anti-Vaccine Movement Threatens Us All,* which was selected by *Kirkus Reviews* and *Booklist* as one of the best nonfiction books of 2011; *Do You Believe in Magic?: The Sense and Nonsense of Alternative Medicine,* which won the Robert P. Balles Prize in Critical Thinking from the Committee for Skeptical Inquiry and was deemed one of the best books of 2013 by National Public Radio; and *Bad Faith: When Religious Belief Undermines Modern Medicine,* selected by the *New York Times Book Review* as an Editor's Choice book in April 2015. Dr. Offit lives in Philadelphia.

# INDEX

## I

## M

## R



## S

## T



## U



## V